a brief history of tomorrow

BY THE SAME AUTHOR

Uri Geller: Magician or Mystic?
Cleese Encounters
The Big Yin
Lenny Henry
Bernard Manning
Michael Palin
Hothouse People
(with Jane Walmsley)

a brief history of tomorrow

The Future, Past and Present

JONATHAN MARGOLIS

2000

BLOOMSBURY

Copyright © 2000 by Jonathan Margolis

Every reasonable effort has been made to trace all copyright holders
of material reproduced in this book, but if any have inadvertently
been overlooked the publishers would be glad to hear from them

Published by Bloomsbury Publishing, New York and London.
Distributed to the trade by St. Martin's Pres

A CIP catalogue record for this book
is available from the Library of Congress

ISBN 1-58234-108-7

First U.S. Edition 2000
10 9 8 7 6 5 4 3 2 1

Typeset by Hewer Text Ltd, Scotland
Printed in England by Clays Ltd, St Ives Plc

To the Future

CONTENTS

The Stud

Look well to this day
For it is life
The very best of life
In its brief course lie all
The realities and truths of existence

The joy of growth
The splendour of action
The glory of power

For yesterday is but a memory
And tomorrow is only a vision
But today, if well lived, makes
Every yesterday a memory of happiness
And every tomorrow a vision of hope
Look well therefore to this day
                    (Sanskrit poem, anon.)

In a studiedly eccentric office just off Baker Street in London's West End – the kind of brightly-coloured, post-minimalist workplace where no one has their own desk, but sets up their laptop wherever they like, where staff scribble their latest brilliant thoughts on wipe-clean wall boards, the office dividers are feng shui waterfalls and there are plates of jelly beans to graze on – a casually-attired group of young brains spends each day thinking about things to come.

There would be no point having an 'Imaginarium', as this place is called by Orange, the British mobile phone company which has set it up right next to its chairman's office, to consider the boring old stuff everyday think tanks ponder on – trends in consumer spending and the like. No, Orange's Imaginarium is a place for more ambitious wheel-spinning on the future of technology and the way it might impact on *homo sapiens*' deep future, far beyond 2001.

Here, for example, is a picture the Imaginarium has painted of the mobile phone some time from now. It's an era when the very concept of making a 'phone call' will have become laughable. Telecommunications will be integrated into our bodies. What used to be simply your mobile phone company, but is now more of a full-time, electronic personal assistant service, will have equipped subscribers with 'The Stud', a tiny device that looks like a small studded earring, but is actually a twenty-four-hour all-purpose link to the rest of the world, just like the one Lieutenant Uhura wore in *Star Trek*.

The Stud will be audible in your ear in a voice and personality of your choosing. It will wake you in the morning, softly calling your name. As you shower – the stud having set the water temperature just so and remotely switched on the coffee maker in the kitchen – it will give a rundown on your schedule, the weather, news items which might interest you, any e-mails or voice messages it has received overnight and whether it would be advisable to drive to the office or take the train today.

On the way to work, it will set up multi-national conference calls, in foreign languages if necessary, translating your words into whatever, and whatever back into your own language. During the day, it will keep tabs on your state of health as well as your car's and that of your finances, guide you if you get lost, tell you where the nearest parking space is, give position and status updates on your children and quietly shop around for the latest grocery bargains which it knows you need.

Naturally, it will connect you at all times to a vastly improved internet. Database information such as sales statistics, spread-sheets, memoranda, and tracking of shipments, or research information such as the GNP of Greece or the projected population increases and purchasing habits of Brazilians, are just a whisper away. Oh, and when you want to 'make a call' to a specific person, The Stud will be enabled for 3-D holographic imaging, meaning you will be able to speak to colleagues, friends and family anywhere in the world, as they appear to be sitting right in front of you as a hologram.

As you see, Orange's imagineers deal in pretty wacky futuristic stuff, not that different from science fiction, or from the American cartoon series, *Futurama*, about a pizza delivery boy accidentally frozen while delivering to a cryogenics lab on New Year's Eve 1999 and not thawed out until 3000, when he emerges into a world of flying cars, Stop 'n'-Drop suicide booths and coin-operated robot hookers.

One small point, however. 'The Stud' is not a scenario for the twenty-fifth or the thirtieth century. It is a real project Orange are working on. They believe it could be available by 2005 – apart, that is, from the holographic imaging. That could take another five years.

What is more, aspects of The Stud project may already be a little out of date. The cybernetics guru Professor Kevin Warwick of Reading University in England is planning personally to test a human telepathy chip some time in 2001. He already has one chip embedded in his body which opens doors as he glides around his department and enables his staff to track him. Soon he will have a half-inch-long device implanted in his arm and wired to a nerve to transmit thought and feeling wordlessly to another, similarly kitted-out human being.

Suddenly, one of the objections to Orange's Stud – that it will turn us into a troop of demented chimpanzees chattering away day and night to an apparently invisible friend – diminishes. Imagine the difference if we only had to *think* to speak to our virtual assistant.

And at the same time, the greatest difficulty with speculating on the future becomes abundantly clear – the increasing speed at which the future is unveiling itself. The famous prediction of computer chip manufacturer Intel's Gordon Moore that the speed and complexity of computers will double every eighteen months, and the machines thus become over a hundred times faster every decade, is turning out to be true. And that's not all, there is also every sign that not only is the rate of discovery accelerating, but that the acceleration rate itself accelerates – with dizzying consequences for humankind.

Think, for example, about this book you are holding in your hands. Now, we have heard the death of print predicted unceasingly over the past few decades. Sony have been developing CD-based electronic book machines since the 1980s without setting off much desire on the part of the public to own one. Instead, we got Amazon.com and a boom in book buying.

Yet, according to Daniel Okrent, Editor-at-Large at *TIME* magazine and a former Editor of New Media at the company, print really is fast approaching the end of its existence. In twenty, or at the outside, forty years, he believes, printed pages will be as quaint as the horse and carriage. In a speech to the Journalism School at Columbia University in 1999, he gave a lucid description of the future of the printed word.

'Imagine this,' Okrent said, 'and if you find it hard to imagine, trust me: I've seen it already, in the development office of a well-established Japanese computer electronics company. Imagine a tablet, maybe half an inch thick, shaped when held one way like an open book or magazine, when turned sideways much like a single page of a newspaper. It weighs six ounces. It's somewhat flexible, which makes it easy to transport. (The truly flexible one, which you'll be able to roll up and put in your pocket, is still a couple of years away, so this one will have to do.) Its screen, utterly glare-free, neither flickers nor fades nor grows dull. To move beyond the first screen in whatever it is that you're reading, you run your finger across the top of the tablet *like this* – a physical metaphor for the turning of the page.

'You are sitting on a beach on a Saturday afternoon with this little wonder, and you're reading this week's *TIME* magazine. Then you decide you'd like something a little more, oh, entertaining. You press a series of buttons *here*, and a cellular hookup to a satellite-connected database instantaneously delivers you

Evelyn Waugh's *Scoop*. And when you've had enough of that –
click, click – you move on, to the football news, or the office
memoranda you didn't finish reading on Friday afternoon, or
whatever it is that you want. Click, click again: each download,
coming to you at dazzling speeds, and a central rights-clearance
computer charges your account, much like a telephone account,
for what you've read or listened to. The satellite operator keeps a
small portion of the income, and the rest goes to the "publisher"
– that is, to the agency that either created the material you are
reading, or that represents the interests of those who created it.
Or imagine this: another message comes to you, from – let's say
Coca Cola. It's an advertising message, and you have been paid
to read it. You have been targeted by Coca Cola, the marketers
from that company have found you on the beach, and for the
privilege of getting their message in front of you they have paid
the satellite operator a carriage fee. The satellite operator,
wanting to guarantee the advertising agency that the impression
has been made, credits your master account a few cents. For
reading the one-minute message from Coca Cola, you get the
first five minutes of tomorrow's electronic newspaper for free.
Everyone's happy.'

Would this not mean the death of literacy, Okrent, a publisher
before he went to *TIME*, went on to ask? Most definitely not, he
concluded. 'My colleagues and I did not grow up wanting to be
in the ink and paper and staples business; we wanted to be in –
we *are* in – the business of words and sentences and pictures and
ideas. Don't worry about the future of newspapers or magazines
or books any more than you would worry about corrugated
boxes or shrink-wrap. They are containers; the substance resides
elsewhere.'

With a rate of technological progress far greater than exponential, it is no wonder that predicting the future ten, a hundred,
a thousand, or occasionally a million years ahead, as I shall
attempt to do in this book, is such a risky business. Samuel
Goldwyn was more accurate than he knew when he coined the
maxim, 'Never predict anything – especially the future'. For even
next week can spring surprises.

In 1997, for example, a retired London University professor of experimental physics, John Hasted, predicted to me over tea at his house in Cornwall that conventional transport would one day become obsolete and be replaced by instantaneous teleportation à-la-*Star Trek*. He warned that this wouldn't happen in his lifetime or even mine. Yet, within a week of our discussion, a paper was published in *Nature* magazine describing the first ever successful teleportation experiment by a team in Innsbruck, Austria. They had succeeded in transporting a subatomic particle – more correctly, a state – from one side of a laboratory to another without any sign of it having travelled through the intervening space. It was some way from 'Beam me up, Scottie', but a scientific sensation nonetheless, as it provided evidence that the mysterious principles of quantum mechanics, the strange, counter-intuitive way in which scientists believe matter behaves at its smallest, most basic, level, might work in practice, and not just on the blackboard of the theoretical physicist.

I have been fascinated by the future since I started writing. The first piece I had published outside of a school magazine, in *The Spectator* in 1972, when I was 17, was a pastiche of George Orwell's *1984*. It was a pretty juvenile attempt to anticipate a future Britain in which it was illegal *not* to dissent from the diktats of a political paradox I had dreamed up – an authoritarian liberal government. I think it was meant as a kind of a warning against the knee-jerk liberalism of the hippy era. If only I'd extended the thought a little more subtly and envisioned a world where you didn't have to *be* liberal and 'right on', so long as you used the empowering *language* of liberalism and sounded the part, I might have been the first person to discover political correctness. But, naturally, I wasn't.

It's that elusive character of the future that I find so enchanting, the way you can get so tantalizingly close to anticipating things, but then catastrophically forget to factor in some key element which evolves by surprise, like political correctness, or cellular phones, or the collapse of the Soviet Union.

I became particularly interested in how, at a personal and

business level, a minority of people have a flair for anticipating the future, while most of us, for reasons of caution, lesser intelligence or lack of imagination, just can't see things coming. This general inability to understand how stuff tends to work out manifests itself at the micro level in the way most of us unthinkingly join the longest line at the post office, and the fact that the most popular lottery number combination is said to be 1,2,3,4,5,6 – no less possible than any other combination, but, well, you know . . .

Central to the problem of anticipating the future seems to be a widespread human failing to appreciate that everything new sooner or later (but generally sooner) becomes old. Yet how many successful businesses do we see floundering not because of bad luck or bad product, but simply because they become complacent and refuse to accept that times and tastes change?

It is remarkable, nevertheless, just how many people, both in business and in futurology, anticipate things correctly. Whether it's down to flair, doggedness or playing hunches, entrepreneurs like Richard Branson or Jeff Bezos of Amazon.com are above all talented futurologists. When it comes to people who have written about the future over the past few hundred years, it is similarly hard not to be impressed by how close, yet simultaneously wide of the mark, they have frequently been.

I returned to the futurological theme several times when I became a professional journalist, examining both the successes of past futurology and its more amusing failures in my articles and columns for the London *Evening Standard* and, latterly, the *Financial Times*.

Although I attempt to be sensibly sceptical of all things modern and futuristic both in those columns and in the book you are about to read, I ought to point out that I am at the same time a passionate believer in the future. I was brought up on stories of the depressed 1930s in Britain, and always warned to be intensely suspicious of the phrase 'the good old days'. As a result, I can't wait to see the future, and am extremely optimistic about much of it.

Today, I regard the internet as the most wonderful human

creation since printing, and believe the most inspiring place on the planet (well, equal first with Venice, anyway) is the Millennium Mile on the south bank of London's River Thames. From the giant ferris wheel opposite the Houses of Parliament to the National Theatre, to the most awe-inspiring art gallery in the world, the Tate Modern, to the reconstructed Globe Theatre, to the little house where Sir Christopher Wren lived during the construction of the preposterously futuristic St Paul's Cathedral directly across the river, to the beautiful, gleaming new Millennium footbridge, to the barely modernized, Dickensian Borough fruit and vegetable Market. This breathtaking sweep of culture, architecture, history and modernity is, for me, the nearest thing I know to stepping into the future I've always longed for.

This book was triggered by a conversation in the spring of 1999 with my editor at the *FT*, Gillian de Bono, in the pleasingly futuristic restaurant Fish!, which stands incongruously in the heart of that strange old produce market. We were talking about the developing media frenzy in the run-up to the year 2000.

Gillian asked if it might be possible to jump the journalistic gun by a millennium or so and write an article speculating instead on the prospects for *3000*. Peering as far ahead as 3000 was a perplexing task for the futurologists I spoke to. They had to imagine a day when developments we can still barely conceive of or would be able to understand are already as distant and remote an historical legend as the days of Aethelred the Unready, the reigning English monarch in the year 1000.

The period around 2015, when computers became as clever as people, will be part of 'early medieval' history by 3000. A similarly ancient legend will be the moment in 2020 when a single desktop computer was developed that was as powerful as all the machines in California's Silicon Valley put together had been twenty years previously; or the moment in 2040, when a computer went on sale 1,000 times more intelligent than Einstein for $1,000 . . .

The resulting article whetted my appetite to explore much further and write a book on the latest thinking about mankind's future for the next century, and sometimes the next millennium, and how it might be viewed in the light of our past attempts at futurology.

It is to Gillian de Bono, therefore, that I am indebted for the idea for *A Brief History of Tomorrow*. Equally important has been my friend Professor Marcello Truzzi, a sociologist at Eastern Michigan University at Ann Arbor, whose astonishingly wide-ranging reading on the future (plus just about everything else) has helped me hack my way out of more journalistic thickets than I'd care to admit to. Graeme Gourlay, who is equally knowledgeable about almost everything, was also enormously helpful, as was Michael McCarthy of the *Independent*, Professor Nick Webster, of the University of California, San Diego, Bill Swainson and Edward Faulkner at Bloomsbury for their continual editorial help and guidance and – as ever – my research assistant, Gabrielle Morris, and agent Vivienne Schuster.

I'd also like to thank: Professor Murray Gell-Mann of the Santa Fe Institute in New Mexico, Kenny Hirschhorn of Orange Plc, Daniel Okrent of *TIME* magazine in New York, Graham May of Leeds Metropolitan University, Professor John Hasted, Ingrid Lander of the University of Technology, Sydney, Dr Rick Slaughter of the Australian Foresight Institute in Melbourne, Simon Craven of British Telecom's research laboratories' press department, Catriona Kelly of New College, Oxford, Moshe Dror of the Israel Futures Society, Dr Ikram Azam of the Pakistan Futuristics Foundation and Institute, Rei Uda Kawashima of the Creative Futurists' Node in Tokyo, Professor John Erickson of Edinburgh University, Dr Bill McGuire of University College, London, Kevin E. Trenberth of the National Center for Atmospheric Research in Colorado, Professor Christopher Wills of the University of California, San Diego, Dr Eric Wolff of the British Antarctic Survey, Cambridge, Dr Alan Dixson of San Diego Zoo, Dr Tim Boon of the Science Museum in London, James Halperin, Dr Vincent Walsh of Oxford University's Department of Experimental Psychology, Ray Kurzweil, Steve Burwen of Intel, Marc Demarest of DecisionPoint Applications, Portland, Dr Michael Argyle of Oxford University, Nico Stehr of the Institute of Hydrophysics in Geesthacht, Germany, Dr Brian

Martin of the University of Wollongong, NSW, Professor Henry Bauer at Virginia Polytechnic Institute and State University, Tony Edwards, Dr Mae-Wan Ho of the Open University in Milton Keynes, Andre Jordan, Dr Shmuel Halevi, Hannah Shepherd, Professor Rodney Howes of South Bank University, Roy and Vera Stewart, Nancy Slessenger of Vine House Essential, David Stewart of Perekhid Media Group – and Penny Beadle, for the beautifully apt Sanskrit poem at the start of this preface.

However, as it's always important to add on these occasions – the mistakes, and I know there will be plenty of them, even if they haven't happened yet, are all my own.

Jonathan Margolis, London, June 2000

ACKNOWLEDGEMENTS

Grateful acknowledgement is made to the following for permission to reprint previously published material: Weidenfeld & Nicolson for permission to quote from *The Age of Automation* by Leon Bagrit; Yale University Press for permission to quote from *Ecology in the Twentieth Century: A History* by Anna Bramwell; Victor Gollancz for permission to quote from *The Mighty Micro: The Impact of the Micro-Chip Revolution* by Christopher Evans; Bill Hamilton as the Literary Executor of the Estate of the Late Sonia Brownell Orwell, Martin Secker & Warburg Ltd for permission to quote from *Nineteen Eighty-Four* copyright © George Orwell, 1949; The Telegraph Group Ltd for permission to quote Stephen Hawking interviewed by Nigel Farndale.

Chapter 1

THE WAY WE WEREN'T

Future Past

'The future,' science fiction guru Arthur C. Clarke once said, 'isn't what it used to be.' A clever, ironic statement, superficially quite ridiculous of course, containing nevertheless two nuggets of wisdom for the price of one. Because what the author of *2001: A Space Odyssey* was talking about, naturally, was futurology, the fusion of informed scientific analysis and inspired guesswork about the future at which he has excelled for over fifty years.

And what he was pointing out was the twin paradox which in the last weeks of 1999 started me off, a journalist of late twentieth-century vintage with ambitions of seeing out the first gasp of the twenty-first, on an inquiry into whether futurology has a future – or should finally be written off by we wise twenty-first-century-ites as so much crystal balls.

For futurology, according to the man who foresaw communications satellites when milk was still being delivered in Britain by horse and cart, has always had two inherent flaws.

Firstly, visions of the distant future tend to be shaped and coloured by the experiences and prejudices of the present; they go out of date and look and sound old fashioned long before the time they were trying to predict comes round.

The second, parallel, meaning of Clarke's observation (which has also been made in similar words by the French poet, Paul Valéry) was more personal for him. In many ways, he has been the wisest and most accurate of futurologists. He is the writer who gave the year 2001 its special resonance as a metaphor for deep space travel, just as George Orwell made the year 1984 a byword for the nightmare authoritarian state. But as Clarke looks around him at the reality of the turn of the millennium, he concedes that it hasn't turned out exactly as he imagined.

In some way, he feels, the future was not supposed to happen so quickly. Yet at the same time, there's no manned mission to Jupiter, no broody, jealous computer with brittle feelings to be hurt, no evidence of aliens having visited Earth, and even Pan Am, the airline featured in the film of *2001* whisking the astronauts to the space station for the start of their flight to Jupiter, has shrunk down to a tiny local carrier in New Hampshire.

Which is why if there's one thing Clarke can be sure of, it is that he can't be sure about the future. And why Massachusetts Institute of Technology professor Noam Chomsky believes, 'Perhaps the most plausible prediction is that any prediction about serious matters is likely to be off the mark except by accident.' And equally why, in the first week of January 2000, when I sat down to start mulling over this first chapter of *A Brief History of Tomorrow*, I was already well aware of what a slippery character we're dealing with when we try to grapple with the future.

Talking to futurologists and reading their predictions through the ages was one thing, but it was the happy coincidence of where I happened to be that morning, when I started the final phase of planning this book, that drove home my awareness of the future's capricious nature.

The twenty-first century had arrived, but you only had to walk down the street to see that things were pretty much the same as they were the previous week, back in the second millennium. Pictures of Princess Diana were still being reproduced in newspapers and magazines at every opportunity, despite her having been dead almost three years. The wheezing old railways in Britain were still using rolling stock built in the 1960s. Old men were still bafflingly wearing hats, scarves and gloves to drive, as if unable to believe cars had come with heaters as standard for at least thirty years. And those cars were still powered by fossil fuels, a mulch of ancient dead leaves and animals, which sounds like such an inspiringly ecological technology – were the opposite not so tragically the case.

More disappointingly in this January 2000 suburban panorama, there were no flying cars, no sparkly Lurex space suits, no orgasmatrons, nobody wearing strange metallic hats which doubled as space aerials, and absolutely no sign of those severe permanent frowns that had for a century been depicted as what would be the facial expression of choice come the year 2000. So much for futurology.

But then again, perhaps the future *had* arrived as advertised, but had done so by stealth, incrementally, and over such an extended period that it wasn't obvious. With this thought in mind, I wandered into a new Starbucks coffee shop that had just opened around the corner in west London.

And after a few moments there – despite the absence of Labi Siffre actually being on hand to sing 'I Can See Clearly Now' – I began to wonder if the real future was not after all going to consist of gadgets or the internet or genetic engineering or Mars colonies. Not entirely, anyway. The real 2000, the most eagerly anticipated year there has ever been, the symbolic gateway to the future for these past several hundred years, was right here, under my nose, waking up and smelling of coffee.

A Starbucks coffee shop in a British suburb on the first working, weekday morning of 2000 was far from being merely an ephemeral, pretentious colonial outpost of a passing trend imported from the United States, an attempt to make everyone in the world feel for the price of a latte that they are one of the friends in *Friends* or an amused acquaintance of the Crane brothers in *Frasier*.

Here, rather, was a paradigm for both the potential and the limitations of futurology, a template not for the way things will necessarily *be* fifty, one hundred or a thousand years from now, but for the kind of wised-up, multidisciplinary thinking futurologists must use – and, crucially, *are* using – to create a model of the future as it is most likely to work out.

Just as Clarke's *2001* and Orwell's *1984* were wrong in

particulars and timing but perceptive and accurate in spirit and much detail, a Starbucks in early twenty-first-century Europe and the things its customers were carrying, wearing and doing, encompassed elements of the knowable, but totally unforeseen, elements of the knowable and thoroughly predicted – and elements, too, of the utterly fickle and unpredictable.

Before you say, 'But hey, it's just a coffee bar,' let me explain how it would seem to a futurologist from two or three hundred years ago, when speculating seriously about the future first caught on.

In Starbucks we have a company which grew out of a single 1970s coffee stall in the Fisherman's Market in Seattle; so it comes to us in a cloned, globalized form from a part of the world which was barely even inhabited a couple of centuries ago, and is now the world's leading centre for two businesses which didn't exist then either – aircraft manufacture and computer software – plus one which very much did, namely coffee houses.

So it's an American coffee house serving an Arabic drink in many forms, some a little outlandish, most with Italian names. The enormous range of varieties of coffee on sale and the combinations available tells us much about the twentieth-century desire to cater to consumers who want everything supplied instantly to their personal preferences. (Aldous Huxley clearly got part of the future right when he wrote *Brave New World* in 1931 and had Mustapha Mond, one of the Ten World Controllers, ask, 'Has any of you been compelled to live through a long time-interval between the consciousness of desire and its fulfilment?')

But at the same time as pandering to the spoilt consumer, the company running the chain prides itself on having a lofty mission beyond the mere selling of cups of coffee. It sponsors programmes for charitable giving, environmental causes and literacy projects; it attempts to educate the public on coffee in general. Nevertheless, it's still a red-in-tooth capitalist venture. When all's said and done, it would far rather we went to Starbucks to drink coffee than to a competitor, and, lest we forget, Starbucks was one of the key companies identified by anti-capitalism rioters in Seattle in 2000 as a target for trashing.

But the most significant thing about my local Starbucks branch was that it is there at all. Fashionable people in Seattle began to appreciate twenty-five years ago having a 'third place', a neighbourhood hangout which was neither work nor home – a distant American cousin, perhaps, of the Parisian left bank café. The British futurology think-tank, the Henley Centre, has predicted the rise of what it calls 'the domestication of leisure' – meaning the development of restaurants and pubs that seem more like home than 'out'.

But the fact that such a business has now set up shop in a stuffy English suburb, where just twenty years ago the choice was stale coffee, stewed tea and nothing much else, shows clearly how American ways – the combination of trendified capitalism with keeping the increasingly demanding, egoistic customer satisfied – have by 2000 become the world's ways. This Starbucks happened to be in west London, but it could have been one of the eight branches the chain has in Beijing, the ten in Bangkok or the one in Beirut.

A coffee shop may be a coffee shop, but it is manifest that the world over consumers cleave above all to brands. The American world, which in so many ways resembles the Roman Empire, routinely squashes historical sensibilities flat. The ancient British middle class wasted no time taking to togas; and the modern, urban British feel as secure in a funky American-branded Starbucks as in a tea shop with paper doilies.

There are more striking examples still of the brand power of what might be called Pax Americana. Journalist Tom Friedman of the *New York Times* has even suggested that such pan-global American companies are a force for world peace. His 'Golden Arches Theory of Conflict Prevention' suggests that no two nations with a McDonald's are likely to go to war with one another. Although strictly speaking, NATO's bombing of Yugoslavia in 1999 contravened that rule, Friedman's law still pretty much holds. It is probably more significant that less than thirty years after the end of the Vietnam war, Microsoft maintains an eleven-person office in Hanoi. When people want Windows, it seems, they are prepared to forget wars and politics.

Yet even if Starbucks is an unashamedly American entity, the interior of its shops are an artfully contrived kaleidoscopic national mix – no overt, one-culture Coca-Colonization here. The shop's name, for example, has a distinctly Olde English air (albeit with a tinge of nautical American via *Moby Dick*). Its logo depicts some kind of Lady Godiva character, with a bit of a come-on in her secret smile. But the stripped wood flooring is definitely Danish in influence (even if we were supposed to be living in an entirely washable PVC world by 2000), while the music playing through the Japanese hi-fi system is African and Latin American. Japan, a nation still primitive and mysterious to our ancient futurologist, also devises and makes most of the modern tools the staff use, from the electronic cash tills to their own wristwatches.

While the natural look reigns at floor level, the coffee house's ceiling is a network of undisguised factory piping, possibly false, suggesting a modern, forward-looking world where we are friends with industrialization and even believe it to be aesthetically attractive. The furniture, by contrast, is not very modern at all, a mixture of natural wood and the kind of brown velour sofas our parents threw out – a suggestion of reverence for the past, here. There's evidence too that these twenty-first-century folk are deep thinkers – one wall of the coffee house is given over to a colourful mural of handwritten philosophical musings. ('It is only when they stop growing that humans become old,' reads a typical one.)

The Starbucks customers seem to be from all over the world, speaking English in a variety of accents, plus several foreign tongues. Puzzlingly for the futurologist of the past, there is also a seemingly random, and quite scandalous, mixing of races, with no effort made to provide separate areas for those of different skin colour. There is even the visual and conceptual shock of white members of staff serving black customers.

There's something still more bewildering, too. Although there is not a single man wearing a necktie – evidence that this is a place of leisure rather than a successor to the eighteenth-century English coffee house, where the coffee-drinking was incidental to the men doing business – some of the men here are clearly *working*, frowning in concentration as they leaf through files of documents, highlighting points, scribbling notes. The fact that they are working in a leisure setting is a bit surprising because it was always imagined that, by 2000, the working week would have withered down to a few hours, while these men's labours indicate that people may be working harder than ever. But their use of paper and pen suggests that the majority of brainwork is still done in 2000 with the aid of traditional tools. Except that most of the men working as they drink their coffee also have beside them technological tools which would have been dismissed as a preposterous fantasy as late as 1980.

They all sport pocket personal communicators, with which they can speak in an instant to more or less anyone else in the world. Most of the businessmen moreover have pocket computers which house in their memory libraries full of instantly accessible information, as well as performing dazzling calculations in a fraction of a second which would once have taken hours. What is more, these pocket computers can be linked through the pocket communicators – and all the while still sitting on a sofa in Starbucks – to a global brain, a universal, shared consciousness, from which can be summoned up in a few seconds in readable or viewable form practically any fact, opinion, news, data, theory, gossip, work of art, shopping offer or joke in existence anywhere on the planet.

Oh, and there's one last surprise, too. These multiracial latte and cappuccino drinkers, mostly in their thirties and forties, aren't all men.

Many are women, who would have been considered old crones three hundred years ago, but have beside them their own babies. Some of the mothers, more puzzlingly still, seem to be engaged in easy, social conversation, meeting on equal terms, with men who clearly are not their husbands. Over half of them, furthermore, are wearing trousers. Some of the women are even working, just like the businessmen.

At the turn of the twentieth century, women were not considered (at least by men) to be mentally capable of even voting. As late as the 1950s a home economics textbook advised women: 'Don't ask [your husband] questions about his actions or question his judgment or integrity. Remember, he is the master of the house and as such will exercise his fairness and truthfulness. You have no right to question him. A good wife always knows her place.'

A Starbucks coffee shop is, then, an object lesson in unpredictability. Just to emphasize the point, while I was there that morning, a British Airways Concorde flew low over the area on its approach to Heathrow Airport; its terrifying noise stopped conversations for a few moments. Design wise, it still looks a little like 'the future', even though the plans for it were originally drawn up in 1957. Yet by 2000, it was regarded as an anachronism – not because it is too slow or old fashioned, but because, despite its supersonic capability, unmatched before or since, it is too noisy, uneconomical, polluting – and possibly even unsafe – for modern use. And this even though it is powered by that same, wonderfully green-sounding stew of dead flora and fauna as cars. Concorde provides another object lesson in futurology – a development futuristic in its day but predicated on the wrong future.

Would it have been possible, then, for a nineteenth-century or a 1950s futurologist to have predicted that by the fateful year 2000, Concorde would be as dead-end a technology as the airship, whereas Dan Dare-type pocket communicators would be infinitely more sophisticated than anyone ever envisaged? Or that these gadgets would typically be used in that distant nirvana, where work was expected to have all but disappeared, by workaholic men and women sitting in retro, natural wood-themed coffee bars run as fully capitalist enterprises but underwritten by a hippy-esque, caring-sharing ethic?

Well, perhaps it could all have been forecast. Everything, after all, that makes our modern world was in existence in the 1850s or the 1950s, just as cave-dwellers could theoretically have built an iMac. In making our cellular WAP phones and palmtop computers, we don't have the benefit of strange new minerals imported from Mars or arcane intelligence taught us by aliens from Vega. So why didn't futurologists of the past ever seem to get the future right? And are we at all justified in imagining that we in the early twenty-first century can finally see into our future any more clearly than they did?

My exploration of the current state of futurology started, I admit, with the belief that it always was and always will be a comedy of errors. We all enjoy with smug hindsight the utterances of such men as the Roman engineer Sextus Julius Frontinus, who declared: 'Inventions have long since reached their limit, and I see no hope for further development.'

This kind of amusingly blinkered thinking was not limited to Roman times. There was a serious proposal in 1899 to save money by closing down the US Patent Office on the grounds that, as its commissioner, Charles Duell, said, 'Everything that can be invented has been invented.' And even if new inventions came along, there were plenty of people prepared to turn a blind eye to them. In 1909 a senior British customs official made a last stand worthy of General Custer against the preposterous invention of the aeroplane. Louis Blériot had successfully flown across the English Channel that year, and it was time for bureaucracy to limp into action. HM Customs decreed, however, that officials should simply ignore cheeky so-called travellers who arrived in the UK by air, since to do otherwise 'would only bring the department into ridicule'.

Even brilliant people who should know better have demonstrated a penchant for shortsighted prediction, especially in their own field. The Nobel Prize-winning physicist, Ernest Rutherford, the founder of nuclear physics, once declared that talk of nuclear power was 'moonshine'. Britain's Astronomer Royal, Sir Harold Spencer Jones, dismissed the idea of space flight as 'bunk' in 1957 – a timely fortnight before the Soviet Union launched Sputnik 1. And that's not to forget Thomas J. Watson, the former CEO of IBM, who never quite lived down a statement he made in the late 1940s: 'I think there is a world market for maybe five computers,' he said. (Perhaps he really meant five computers *per household*. I think we have about seven including laptops and palmtops, plus four or five redundant ones in the attic, but then I don't pretend to be really into computers.)

No wonder that, in another of his dicta, Arthur C. Clarke said that 'when a distinguished but elderly scientist states that something is possible, he is almost certainly right, but when he states that something is impossible, he is very probably wrong.'

Decades of futuristic films, comic strips and science fiction have done little more to help the cause of futurology, or so it seems. Their main contribution has been to demonstrate repeatedly that we would always be imprisoned in the present, with an 1890s or a 1930s future invariably looking nothing more than a slightly streamlined, stylized version of the 1890s or the 1930s.

To suppose, then, that we today could somehow avoid those same amusing pitfalls seemed to be a pitfall in itself, an example of what C.S. Lewis described as 'the snobbery of chronology' and which I shall call 'the arrogance of the present' – the belief of every successive generation that at last, sophisticated, modern folk that we are, *We've Got It*, and indeed, *We ARE it*.

One of the best comments on our present-centric attitude to the future is by the great cartoonist – from Seattle, as it happens – Gary Larson. A 1996 cartoon of his depicts a scene at a caveman-era future theme park, Future Werld, a sort of Neolithic EPCOT, in which a cave person in an animal skin is seen glowering at a huge picture of a slightly more evolved cave person who is beaming as he strikes a match. The present caveman is sneering at the picture and snorting, 'Yeah, right.'

The arrogance of the present, the belief of every generation that it alone has been chosen to live in a 'special' time – even if that means an especially bad time – is perfectly understandable. Princeton astrophysicist Professor Richard Gott has developed a mathematical theory of futurology, the nub of which is that all generations mistakenly believe themselves to be living through such a special time. He explains that it's natural to want to think of your generation being at the beginning of a great epoch, or in an apocalyptic situation at the end (but never *too* close to the

brink, of course) of another. It adds drama to life, Gott argues –
but it's far more likely to be misleading than it is to be true.

The arrogance of the present is the enemy of intuitive,
inspirational forecasting just as it is of cool, measured, analytical
futurology. The first cousin of plain egotism, arrogance of the
present, colours, I would estimate, 95 per cent of what is
routinely spouted in the mass media and politics about the
future.

It corrodes the thinking of even those of us who like to think
we are somehow immune. How often do we read about some
new technology, sigh and put down the newspaper and wonder
if science has at long last reached its terminal velocity, just as the
Victorians reputedly did when they calculated that trains could
never travel at more than 40 mph because the human frame
could not survive anything faster?

Drive down London's Great West Road past the British head
office of Gillette, the razor blade makers, any night and you will
see dozens of windows lit with figures behind them sitting at
computer terminals. What are they *doing* in there? Perfecting
some even *more* efficient shaving system with not three blades,
but four, five, more? I like to imagine there being some martinet
foreman bellowing at the rank and file razor blade designers,
'Sharper! Come on! Sharper!' Of course, there isn't; but of
course at any given time, Gillette will be working on a still
better blade. To assume shaving or anything else has finally
reached terminal velocity because it's 2001 and, hey, this is my
epoch, is arrogant to the point of absurdity.

Yet even with the arrogance of the present such an insidious
factor, I have come to believe that routinely disparaging serious
futurology as so much guesswork is unfair and uninformed, and
makes us much like that over-sceptical caveman. For when you
examine their predictions fairly, as I hope I do below, the
futurologists of the past don't seem to have done nearly as
badly as we imagine.

Those of the turn of the nineteenth century, when, helpfully,
there was already a history of futurology to look back on and
learn from, did particularly well. For every comically off-beam

forecast they made, there were several, less celebrated, which have turned out to be spot on. There is a tendency to forget those who make predictions that hit the mark, perhaps they make us feel we have progressed less than we imagine. Yet with the precedent set by the foresight of the more successful early futurologists, I have become increasingly sure that today we can be more confident than ever before of predicting our future to a considerable degree of accuracy.

The consistent mistake made throughout the history of futurology has been to scramble timings; things frequently – normally, indeed – appear as predicted, but rarely at the right time. The other thing which has almost always gone wrong with prediction in the past is more problematic, yet, I would argue, even less important. It is the failure to anticipate the fickle finger of fashion, the elements of the Starbucks factor which are merely aesthetic.

The question is: How significant in the greater scheme of things is the 'look' of a particular era? Fashion pages and style supplements would say they are overridingly so. Yet surely external appearances are something chosen more or less at random rather than as a rational response to need? Futurologists have little choice but to guess at the precise aesthetic of even a couple of years ahead. But it is so much packaging, and cannot be said to matter hugely; what is more, concentrating on how futuristic visions inevitably make a hash of what the future will physically look like detracts from their often impressive substance. And laughing at early visions of the future because our ancestors got the colours or the hem lengths or the degree of flare on the trousers wrong is, when you think of it, really just another example of the arrogance of the present.

The intelligent human being's ability to work out by analysis and intuition what lies ahead for our species is relatively new, for the history of the future has been a remarkably short time developing. Only a few centuries ago, the future barely existed in Western human minds. If it was anyone's business, it was God's, and any attempt to foretell – let alone to influence or shape it –

was impertinence, tantamount to witchcraft. Not that anyone really wanted to do so. Time was vague without clocks; people lived in a static world with no concept of progress; the future was tomorrow's weather, wondering if the crops would fail or if taxes would go up. Beyond that, for the medieval European, was the dour prospect of dying in your thirties at the latest, and most probably joining the eternally damned in Hell.

It was only in the late Middle Ages and early Renaissance that the European intellectual elite started to think in any sense about the future, and then it was only on a need-to-know basis. The idea that we should progress to better things was by no means accepted. Regress was usually more valued. Educated people believed that following the ancient civilizations was the correct way to conduct modern affairs; Machiavelli, for example, who lived from 1469 to 1527, argued that gunpowder should not be used in war because the Romans didn't use it.

Of course, the belief that the subsequent enlightenment, which necessarily incorporated a measure of forward thinking, was an exclusively European phenomenon is a ridiculous European conceit. Certainly, it had an Italian element in that the Renaissance in Italy was the great-grandfather of general European Enlightenment. But the Italian Renaissance itself was in large part derived from imported Chinese science and Indian and Arabic mathematics. And enlightenment as a way of thinking was expressed as an ideal thousands of years before the Renaissance in ancient writings in a variety of languages, from Sanskrit to Hebrew to Pali. Two and a half thousand years ago, indeed, Gautama, the founder of Buddhism, was named 'Buddha', meaning 'the enlightened one'.

The rather late beginnings of science and commerce in Europe had begun, nevertheless, to require a measure of genuine anticipation, whether this was to help predict how long a chemical reaction might take to work, or to create a time limit to a contract. And necessity drove the Europeans to be quite good at thinking ahead.

The existence of rival calendars, a disparity which had never previously mattered a jot, began to seem absurd, so calendars merged; clocks, too, gained popularity once they became a necessity, although for hundreds of years, they were almost too accurate. Local time was taken literally, so clocks in a city like Bristol, a hundred miles west of London, were set nine minutes behind those in the capital. It was only when the coming of railways and the need for rational timetables made coordinating time a necessity that the concept of time zones was introduced.

Such methods of futurology as there were in Machiavelli's day were what could generously be called conjecture, less generously, shameless invention – as well they might, since for hundreds of years, the prevailing view had been that mankind was in a state of terminal deterioration and decline, and was unquestionably reaching the end of its days. So when, in sixteenth-century France, the successful and respected physician Nostradamus weighed in with seven whole volumes foretelling 'the future events of the entire world', he was demonstrating a quite revolutionary optimism merely by suggesting humankind would survive at least until 1999, when he prophesied that 'from the sky there will come a great king of terror'. It was no wonder that Nostradamus was popular; for a prominent scientist today to guarantee that, despite all our seemingly intractable problems, we will survive even another thousand years (if only then to face a sticky end) would also be pretty big news.

Nostradamus's continuing popularity today, even after the disastrous failure of any 'great king of terror' to materialize in 1999, is a stranger business altogether. But it may not be so surprising that millions are still prepared to take a supposed prophet's warnings for the future seriously even when he has been consistently wrong at every count in the past. For when people today long for someone like Nostradamus to be regarded as an accurate predictor of the future, and by doing so to get the buzz of imagining they are first with the news that we are on the brink of an apocalypse, they are quite happy to 're-interpret' his predictions. As a British Member of Parliament, Paul Flynn, wryly commented once, 'Only the future is certain. The past is always changing.'

Back in the sixteenth and seventeenth centuries, the growth of science over superstition (not to forget the development of capitalism, too) heralded a growing belief that human beings might just have a future beyond destruction and damnation, and increasingly confident snippets of futurology appear in the writings of Isaac Newton's time. This extraordinary forecast was made in 1661 by the philosopher Joseph Glanvill, a clergyman and chaplain to Charles II, who clearly liked hedging his intellectual bets, because as well as being a scourge of atheists, he was also a founder member of the Royal Society, one of the first scientific societies in Europe.

'To them that come after us,' Glanvill wrote, 'it may be as ordinary to buy a pair of wings to fly to the remotest regions, as now a pair of boots to ride a journey; and to confer at the distance of the Indies by sympathetic conveyances, may be as usual in the future as by literary correspondence . . . I doubt not posterity will find many things that are now rumours verified into practical realities. It may be that, some ages hence, a voyage to the Southern tracts, yea possibly to the Moon, will not be more strange than one to America . . . the restoration of grey hairs to juvenility and the renewing [of] the exhausted marrow may at length be elicited without a miracle; and the turning of the now comparatively desert world into a paradise may not improbably be effected from late agriculture.'

The book in which this uncannily accurate flight of imagination appeared was not a real work of futurology, but a treatise arguing the case for the then radical idea that scientists should try to explain the physical world by doing experiments, rather than by deferring to what the ancients believed. The first complete book ever published about the future did not appear for another seventy years. It was by an Anglo-Irish clergyman, Samuel Madden, and in spite of having an extremely intriguing title – *Memoirs of the Twentieth Century, Volume 1; Being Original Letters of State under George the Sixth* – it was a terrible flop, which achieved the distinction of becoming a rarity within a week, and not because it had sold out.

A handwritten note in the one known surviving copy, now in

the Bodleian Library in Oxford, reveals that a thousand copies of the book were published on 24 March 1732 – a print run which suggests somebody expected it to be a bestseller. A hundred copies were delivered to the author on the twenty-ninth; with just ten going to booksellers around London. Four days later, the remaining 890 copies were returned to Dr Madden to be destroyed. Rumour had it that he had been warned the book might be suppressed, but the reason for the good doctor's junking of four years' of work in a morning remains a mystery.

Madden wrote in his unsigned preface of being 'the first among historians who leaving the beaten tracts of writing with malice or flattery, the accounts of past actions and times, have dared to enter by the help of an infallible guide, into the dark caverns of futurity, and discover the secrets of ages yet to come.'

A promising start, but the lengthy text which follows is less so. To avoid the charge that he is simply making up his vision of the future as he goes along, Madden explains that the book consists of a bundle of letters from his own five-times great-grandson, handed to him by an obliging angel. So apart from translating the letters into 'the English of these illiterate times' (funny how every age believes that the language is going to the dogs), and editing out some parts which 'should be kept Crown or family secrets', there wasn't a lot for Madden to do.

What the book really is is a back-door critique of the British government of the early eighteenth century. Opening with a letter datelined 'Constantinople, Nov 3 1997', there follows 527 pages of rather dull political wish fulfilment, in which George VI is portrayed as a mighty world emperor.

Fleeting ideas of interest can be found amidst the turgid prose, enough to make it possible to imagine that Madden *was* a forward-looking character rather than just a bad satirist. His 1997 sees the establishment of public granaries in all villages, and a law which makes a woman who isn't necessarily ravished, but whose 'soul is ruined by debauching with flattery', entitled to a third of the offending man's estate. This could kindly be interpreted as an early anticipation of women's rights legislation.

Thirty years later, in 1763, another, more ambitious work of futurology appeared, called *The Reign of George VI*. It's curious that the anonymous writer should have picked the same name for his twentieth-century English king, but perhaps in the time of George III, the concept of a George VI was as futuristic as anyone could cope with, like Elizabeth IX might be today.

The George VI of this book reigns from 1900 to 1925. *The Reign of George VI* was a far more successful and colourfully written publication than Madden's attempt, and even came out in a German edition. Again, it is primarily a work of political and military satire, intended to tell the political leaders of the author's day what they ought to be doing. And again, this George is a conquering hero, who tames the nations whom the author believes will be the bugbears of the early twentieth century – France, of course, and Russia.

Russia (25,000 of whose troops sack Durham in a war of 1900) was an interesting choice. The author sees it as the world's second most powerful nation by 1900, which isn't at all a bad prophecy. It is America, the author also correctly forecasts, which will be the land of promise by 1900.

On the question of transport, the anonymous author of *The Reign of George VI* does rather well. The canal building boom had just begun around the time the book was written, and the author extrapolates that by 1900, a network of canals and rivers would form a superhighway network, which would enable communication between every part of the kingdom. 'Villages grew into towns and towns became cities,' he wrote.

There are a few wildcards, of course. It is envisaged that George VI would find London devoid of 'buildings that did honour to the nation'. So in 1907, he builds a kind of Versailles outside a new city called Stanley, near Uppingham in Rutland, the smallest county in England. The architecture in twentieth-century Stanley is strictly controlled and must conform to the classical ideal. The city includes the Law Courts, Parliament, all government offices and a cathedral far grander than St Peter's in Rome. The surrounding landscape, which is regarded as the most beautiful in the world, 'teems with foreign birds and as many harmless beasts as possible procured to run about the woods'. Artificial mountains crowned with little temples and pinnacles, 'a prodigious quantity of masonry and many cascades tumbling down artificial rocks', furthermore, encircle the whole Stanley downtown district.

All this may seem badly off beam, yet it could be asked whether it is really very different from the kind of idealized, verging on the twee, landscape Britain's current monarch-in-waiting, the future Charles III, often talks about, and has even had constructed on his land in Dorset, southern England, in a new olde worlde village called Poundbury.

Where the year 1900 as depicted in *The Reign of George VI* significantly lacks in vision is in the field of science, which is accorded very little standing or importance. The subject of manufacturing takes up just half a paragraph, and while there are pages praising the young king's patronage of the arts and sciences, not a single mention is made of a specific scientific advance, other than unspecified improvements to the constantly upgraded British war machine. Socially, however, there are some developments, which we would recognize as progress; by 1925, we read, George VI has built twenty orphanages at his own expense, and in partnership with local county authorities, thirty-five hospitals.

Although the earliest futuristic books were English, it was a French dramatist, Louis-Sebastien Mercier, who can really be said to have invented the future. In 1770, he wrote a utopian prophecy called *L'An 2440*, which was an instant worldwide bestseller and continued to be so for the next twenty years. Although it was less technology based than Joseph Glanvill's inspired vision of a century earlier, Mercier touched a nerve, ignited a fascination and optimism about the future that has rarely waned over the following 230 years.

Mercier's twenty-fifth-century dream is almost as worth aspiring to now as it was in 1770. If Sir Isaac Newton could explain the physical world, as he recently had, then some rational law of social motion must be at work in human history, Mercier and his fans believed. And that in turn must mean that with reason and hard work and the understanding of such laws of social science, human society would be set on an ever-improving upwards curve.

So Mercier's is a perfect world of peaceful nations, constitutional leaders, universal education and even racial equality; 'We all view each other as brethren and friends,' Mercier wrote. 'Both the Indian and the Chinese are our fellow-citizens as soon as they tread on our soil. We accustom our children to look on the universe as one family, sheltered under the protection of one common father . . . Men have learned to love and esteem each other.' Even more extraordinarily, modern readers will note, people in 2440 are said by Mercier to make voluntary contributions to their national treasury.

Education is now the state's responsibility. Latin and Greek are made obsolete and replaced by the international languages of English, German, Spanish and Italian. All governmental injustice has disappeared, and there are even laws against ostentation. Modest, natural hairstyles have replaced the powdered wig. Most importantly of all, it is the combined power of humanity and technology that has led to such a peaceful planet. The new, hi-tech canals that the author of *The Reign of George VI* was so excited about have now gone international, and been blasted through the landmasses to link the North Sea to the Mediterranean. Technology has also provided 'all sorts of machines for the relief of man in laborious works, and capable of much more force than those in our time'.

Needless to say – and in what would now be regarded as a huge publicity coup – the book became the first account of an ideal world to fall foul of censorship. In 1778, the King of Spain felt his beard sufficiently singed to ban *L'An 2440* for blasphemy and anarchic tendencies. It also, a little more embarrassingly, you would think (although sales only improved still more), fell victim to the speed of technological advance. For while Mercier was confidently proposing canals as the future of transport, half way through the book's run, out of a clear blue sky – literally – came the Montgolfier brothers, Joseph-Michel and Jacques-Etienne, in their hot air balloons.

Only the 1969 Apollo 11 Moon landing can compare with the impact on the world's imagination of the discovery of flight, as it unfolded in ever more breathtaking instalments throughout the summer and autumn of 1783. First, the test flight on 4 June in the marketplace in Annonay, where the balloon rises into the air about 3,000 feet and stays aloft for ten minutes before landing a mile and a half away. The Montgolfiers then go to Versailles, and repeat the experiment on 19 September with a larger balloon carrying a sheep, a chicken and a duck for an eight-minute, two-mile flight. Then, on 21 November, comes the climactic first manned flight by Jean Pilâtre de Rozier and the Marquis d'Arlandes, who soar over Paris for five miles in twenty-five minutes.

Mercier, the Arthur C. Clarke of his day, watching his forecast for the twenty-fifth century go slightly pear shaped within thirteen years of its publication, was among the awe-struck crowd of 200,000 watching the balloon from the Tuilleries in Paris. 'It was a moment which can never be repeated, the most astounding achievement the science of physics has yet given the world,' he wrote, before, doubtless, dashing home to make a few changes to the latest edition of his monster bestseller.

It had been simple enough for Puck in Shakespeare's *A Midsummer Night's Dream* to announce in lines penned less than two hundred years earlier, 'I'll put a girdle round about the earth in forty minutes', and thus make what seems to be a forecast of jet travel and radio waves. But Puck was (a) a sprite, and (b) a fictional character. The new breed of flesh and blood aeronauts drifting across the skies of Europe in the 1780s was clearly on the threshold of shrinking the world by flying. And so far as speculation about the future was concerned, the genie was properly out of the bottle for humankind.

Futuristic-minded writers all over Europe began to produce countless plays, poems and fantasies about flying. And almost as if the communal human spirit was soaring upwards with the balloonists, Mercier's work was imitated by other utopians after 1783. A play, *The Year 2000*, written in 1789 by Nicolas-Edme Restif de la Bretonne, depicted a just community under a benign king, where lawyers have been made unnecessary. 'In the year 2000, virtue never goes unrewarded,' says the King. A curious feature of this version of 2000 is that marriage partners are chosen by a council of elders, and married couples are kept apart for years at a time to increase their passion.

Such oddities apart, futurologists were getting closer to what we now know to be an accurate vision of their future. And as the nineteenth century approached, they inched closer still. An English demographer and economist, Thomas Robert Malthus, son of a wealthy family from Dorking, Surrey, anonymously published a sensational pamphlet in 1798 – two years after he took holy orders – which for the first time in history sounded the alarm against overpopulation.

In *An Essay on the Principle of Population as it Affects the Future Improvement of Society*, Malthus argued that the fashionable utopian hopes for social progress were misconceived because population will always increase more steeply than the growth of production. In other words, we would shortly run out of space on the planet unless we cut the birth rate drastically.

In stating this, Malthus staked a claim as a pioneer of family planning and even of today's green policies. But he also made himself extraordinarily unpopular. He championed contraception (but only for the poor) and severe workhouse regimes to stand as a warning to anyone intending to become a burden on the parish. He regarded famine with equanimity. As a result, he was described by Karl Marx a few decades later as a 'miserable parson' guilty of spreading a 'vile and infamous doctrine, this repulsive blasphemy against man and nature'.

For over one hundred and fifty years, Malthus was derided as scaremonger, and since he had – typically for a futurologist – muffed the timing of his prediction, he probably was. Yet although the word Malthusian is still used as both an insult and a term of approbation, many modern Malthusians continue stubbornly to insist the 'miserable parson' had a point, and it is only a matter of time before he is proved right – even though, in practice, we persist in managing to produce more food per head as population increases.

In the more mainstream futurology of the nineteenth century, meanwhile, it was machines, belching steam and later fizzing with electricity, which became the focus of forward-thinking minds. Jules Verne, a French stockbroker better known as the author of *Journey to the Centre of the Earth* and *Around the World in 80 Days*, was obsessed with machinery and mechanical contrivances. But this was hardly an eccentric fixation for a man who by the time he died in 1905 had seen the introduction of railways, buses, cars, aeroplanes, electric light, radio, the telephone, sound recording and photography.

One of Verne's less known works is a story about the year 2000, called *An Ideal City*. Alongside absurd touches, such as a jokey prediction that women's dresses would have such long trains that they would need wheels, and that babies would be fed by steam-driven breast-feeding machines, there were ideas which have pretty well come to pass. Verne describes, for example, how musical recitals are sent down a wire from the artist to pianos around the world. Well, MP3 music on the internet hasn't worked out *quite* like that, but it's hardly dissimilar. Other mechanical gimmicks in the book haven't come about, but easily could, and would be very welcome today. One such example Verne dreamed up is the automatically self-cleaning street.

So in love with machinery were the imagineers of the late nineteenth century that they often fell into the trap of envisaging technological advances which were more form than substance, more contraption than useful idea. A particularly popular book similarly set in 2000, but this time in the USA, was called *Looking Backward*, by Edward Bellamy. Published in 1888, it featured a utopian, socialist 2000 in which, as a mechanical expression of the communal spirit of the times, instead of people carrying their own umbrellas, there are municipal canopies lowered over the streets when it rains.

A mechanistic approach to how people would work led Bellamy up another prophetic blind alley, too. In his 2000, nothing new is created unless someone puts in an official request for it to the authorities. And there are democratic fallbacks if those authorities are laggardly. 'Suppose an article not before produced is demanded,' one character explains. 'If the administration doubts the reality of the demand, a popular petition guaranteeing a certain basis of consumption compels it to produce the desired article.' Nobody in Bellamy's 2000, however, puts in an order for Tamagotchis, Walkmans, Posh Spice – or even Coca Cola.

Although Edward Bellamy's book was a huge bestseller, it was an American newspaper columnist in the same year, 1888, who provided the more accurate – uncannily accurate – predictions for 2000. David Goodman Croly, known as 'Sir Oracle', wrote for example: 'Women throughout the world will enjoy increased opportunities and privileges. Along with this new freedom will come social tolerance of sexual conduct formerly condoned only in men. In addition, because of the greater availability of jobs, more women will choose not to have children.'

On transportation and cities, as well as social matters, Croly was also remarkably perceptive. 'Navigation of the air will be the most momentous event in history. It will do away with all uncharted regions on the earth and it will enable people to be highly mobile, spending summers in one place and winters in another . . . New York will continue to expand as a major commercial center. Brooklyn and the surrounding towns will form an integral part of the metropolis, which will be characterized by enormously high buildings. A subway will whisk people from one part of the city to another in a matter of minutes.'

Such visions of the future were the subject of much excited talk at the Chicago World's Fair five years later, in 1893. 'By the end of the next century, great corporations and business interests will be conducted harmoniously on the principle of the employees and workers sharing in the profits,' said a journalist covering the fair for *The Century* magazine, Junius Henri Browne.

Browne's interviewees had clearly psyched him into this optimistic frame of mind. 'By the end of the twentieth century, transportation facilities will have so improved that the orange district of Florida will practically furnish the United States all the oranges that it requires,' predicted Samuel Barton, a financier. 'The average man will live to be 100,' pronounced a theologian,

Thomas De Witt Talmage. 'Three hours will constitute a long day's work by the end of the next century. And this work will liberally furnish infinitely more of the benefits of civilization and the comforts of life than 16 hours' slavish toil will today,' said Mary E. Lease, a prominent social reformer of the day.

But socialism wasn't the only flavour of an ideal world on offer in this future-obsessed era. A scientific-minded playboy, John Jacob Astor, wrote of a capitalist utopia in 2000 in his book, *A Journey to Other Worlds*. Astor was a minor inventor, with patents on file for various marine turbines. His book is filled with technological developments he found it easier to imagine than design – yet which appear well short of far fetched today.

Astor envisages rocket-powered spaceships, solar power, tidally generated electricity and battery-powered aeroplanes. The plot of *A Journey to Other Worlds* concerns a corporation called the Terrestrial Axis Straightening Company, which aims to un-tilt the Earth so as to create universal summer. 'Polar bears will soon have to use artificial ice,' one character notes enthusiastically. It's funny that today, manmade global warming offers the same scenario – only for us, it is a fear, not a hope. Funny too, in a grim way, that this enthusiast for big engineering solutions to transport challenges and for warming up the planet died on the *Titanic* when it struck an iceberg.

Other visionaries also had their inspired moments at the turn of the nineteenth century. In 1895, Konstantin Tsolkovsky, a Russian scientist, predicted spacesuits, liquid oxygen and hydrogen-powered rockets and orbiting space stations. Britain's rival to Jules Verne, H.G. Wells, who in the early twentieth century was to foresee tanks, bombers, intercontinental missiles, atomic-type bombs and microwave ovens, came up with a fresh slant on the future in his 1895 fantasy, *The Time Machine*.

By setting his furthest future visualization in the extravagantly futuristic year of 802701 – the eight thousandth century – Wells gave himself scope to out-future his rivals. For that distant time, he proposed a world where there was total, social polarization between the aristocracy, the Eloi, who are tiny, pale, weedy intellectuals, and the Morlocks, who are what remain of the working class, and are ugly brutes who live underground and emerge to devour the Eloi who live above. The same theme was to be popular right up to the present day, when 'brutes-living-underground' has become almost an entire movie genre.

The Victorian belief that technology essentially *is* the future continued to colour early twentieth-century futurology, and the growing fervour for prognostication naturally sparked increased mass media interest; utopian visions were all very well for intellectuals, but now futurologists could discuss the future in terms of contraptions and conveniences people would soon be able to buy.

At the turn of the century, a journalist called John Elfreth Watkins interviewed 'the wisest and most careful men in our greatest institutions of science and learning' on behalf of the American magazine, *The Ladies' Home Journal*, and by luck or judgement, managed to make a brilliant series of consumer predictions. The future, Watkins reported, would witness air-conditioning, international telephones, colour photography, frozen meals, medicine taken by skin patches, snowmobiles and the tapping of energy from wind, tides and sunlight.

Out-of-season fruit – which surely would be one of the most striking innovations to a Victorian visitor to our times – is (almost) accurately predicted by Watkins's experts: 'Strawberries as large as apples will be eaten by our great-great-grandchildren for their Christmas dinners a hundred years hence,' he wrote. Most amazing of all was this prophecy: 'Persons and things of all kinds will be brought within focus of cameras connected electrically with screens at opposite end of circuits, thousands of miles at a span.'

An unnamed journalist from the London *Daily Mail* also scored a series of bulls' eyes predicting the year 2001 in that newspaper's end-of-century edition on 31 December 1900.

Electric energy would by 2001 revolutionize transport and home and industrial heating, the writer promised, abolishing the scourge of soot and even powering trivial implements such as curling tongs and cigar lighters. Railway speeds would considerably exceed the current ceiling of sixty mph. And by the same year, there would be a Channel Tunnel, which would start a process of effectively shrinking the world. The new wireless system being demonstrated by Marconi would hugely increase that shrinkage, and the inventor was to be taken seriously, the journalist argues, when he promised that a way would be found not only for *voices* to be transmitted wirelessly, but for different wireless signals to cross the same airspace without interfering with one another; it is as well Marconi did find such a way, of course, otherwise it would be possible today only for one cellular phone call, broadcast station or radio-controlled garage door to work at the same moment.

In medicine, the *Daily Mail* article suggested, incurable diseases such as cancer would start to 'lose many of their terrors' by 2001. 'Perhaps it may seem that too rosy a view is taken of some twentieth-century possibilities and probabilities,' the piece concluded. 'But in the bright lexicon of the dawning century, there is no such word as impossible.'

The biggest topic for speculation in the USA around 1900 was equality. A noble aspiration, perhaps, but there was a catch; not one of the newspapers was referring to racial or sexual parity, but to the simple erosion of the gulf between rich and poor. And the 1900 version of social advance was a little different from what we have experienced; for the *Chicago Tribune*, the most liberating thing about 2000 was that selective breeding would make women more beautiful; for the *Manchester Guardian* it was that chaperoning of women would die out and that for women to play the harp – or even go to Paris on their own – would no longer be considered daring.

An interesting idealistic prophecy of 1900 which many today may feel has been semi-fulfilled, with the Western armed forces theoretically, at any rate, engaged only in peacekeeping activities, was made in *The Atlantic Monthly*. The magazine envisaged an exhibition in 2000 where redundant military equipment was displayed to remind a public that has all but forgotten war of how 'much of the ingenuity of our people was formerly devoted to warfare'.

The *Saturday Evening Post* of 1900, speculating on the same theme, imagined the people of 2000 saying, 'War is dying out because men have something else to do. They are engaged in trade.' The *Washington Post* had a seemingly cranky theory that technology might by 2000 create weapons so awesome that they would 'compel universal peace'. Many advocates of the paradoxical benefits of nuclear weapons would claim that this is precisely what happened in the second half of the twentieth century.

Remarkably enough, even the e-commerce revolution of today was foreseen around 1900. The journalist and economist John Maynard Keynes predicted it while still a Cambridge student: 'The inhabitant of London could order by telephone, sipping his morning tea in bed, the various products of the whole earth and reasonably expect their early delivery upon his doorstep; he could at the same moment and by the same means adventure his wealth in the natural resources and new enterprises of any quarter of the world, and share, without exertion or even trouble, in their prospective fruits and advantages.' (Today's devotees of e-commerce will doubtless applaud Keynes's precocious foresight, but regret that the 'delivery upon his doorstep' part has yet to be sorted out.)

An American journalist, William Wallace Cook, meanwhile was writing a book called *A Round Trip to the Year 2000*. Published in 1903, it forecast a world in which mechanical creatures, 'muglugs', take over from humans and send them – us – out to grass in the country. Capitalism has also reached absurd frontiers, with the 'Air Trust' charging for air and the 'Sun Trust' buying the rights to charge citizens for sunshine. Absurd? Well, as we will see later, many artificial intelligence experts argue that the former scenario is almost on our doorstep. And the privatization of public utilities, especially water supply, in the UK is seen by many cynics as the first step towards corporations cashing in on what nature supplies for free.

Jean Marc Cote, a French illustrator, specialized around 1910 in drawings of 'life in the year 2000'. Underwater croquet, robots that shaved men and battery-powered roller skates were among the imagined attractions of our time. A drawing of a school classroom of the year 2000 showed books being fed into a machine and their contents being transmitted into the pupils' brains through suitably ridiculous-looking headphones which hang from a network of wires on the ceiling.

The 1920s magazine *Amazing Stories* ('Extravagant Fiction Today . . . Cold Fact Tomorrow') and many imitators later took up the futurist theme in graphic form. Today's view of the future as not merely science but *gadget*-dominated has its roots in these boys' comics. And that future may in a sense have been influenced a little by its fictional pre-publicity. A series on crime-fighting in 2000, for instance, forecast devices which aren't at all unfamiliar to us. There were tiny recorders strapped to wrists, helicopter pursuit of criminals and bloodhound machines that identified a perpetrator's smell. Substitute smell for DNA, and you have a reasonable preview of the crime-fighting gadgetry of 2001. How many of today's inventions, you have to wonder, were actually prompted by memories of futuristic comics?

In Russia, meanwhile, the Bolsheviks were developing what they profoundly believed was a method of jumpstarting an increasingly backward society and transporting it into a made-to-measure future. The vision of socialist utopias started appearing in Russian literature around 1910. 'At first, this literary futurist movement was highly derivative,' explains Dr Catriona Kelly, Reader in Russian at New College, Oxford. 'Russian futurists picked up on a lot of Italian writing and added local detail. But then some weird Russian futurist ideas grew up, such as a belief that if you could avoid reproduction then somehow you would make a more perfect society. Tolstoy maintained – although he kept changing his views – that all sexual intercourse is sinful, including marital intercourse, so you just ought to avoid it and if the human race dies out then so much the worse for the human race.'

There were more mainstream strands in the utopian movement similar to those in the West at the same time. 'One of them is the movement out of the city, so there was a Soviet garden city movement, which isn't well known,' says Dr Kelly. 'Sometimes these settlements would take the form of little concrete pods out in the countryside, and there were also communes which were about changing individual behaviour and living together in small communities.

'But technology was also very important in these visions. They were especially fond of the idea of the human body as a machine. There was a popular song called "The March of the Aviators" which has this idea of the human heart being a motor and the arms being like propellers. It became incredibly popular during the 1930s and is the sort of thing that everybody knew.'

Elsewhere in Eastern Europe, in Czechoslovakia, another mechanized 'March of the Aviators'-style vision of the future was being hatched by the writer Karel Čapek, whose play *R.U.R.* (Rossum's Universal Robots) envisaged a race of humanoid machines which develop a personality, become romantically involved with other robots, then with humans.

Čapek began a fascination amongst we flesh and blood humans with artificial people. He named the machines robots, after the Czech for hard labour, and meant them to be friendly. Generations of subsequent cinematic robots, nonetheless, from film director James Whale's 1931 take on Frankenstein's monster to the Daleks to the Terminator, have been thoroughly *un*sociable, even if in real life, all attempts so far at building such creatures have been little more than comical. For the most sophisticated robots we have eighty years after Čapek's original flight of fancy cannot begin to approach the level of complexity required to walk up stairs while carrying a cup of tea.

As the 1920s progressed, the new towering urban landscape in the United States (dictated as it happened by high real estate prices in Chicago and New York and the need to build upwards, rather than any great desire to be futuristic) inspired the German director Fritz Lang, after seeing the blazing lights of Manhattan from a ship, to make his 1927 film *Metropolis*.

Lang's was yet another vision of the ubiquitous year 2000, and became a favourite film of Hitler's. Not unlike *The Time Machine*, *Metropolis* glimpsed a world of plutocrats living in luxury while the proletariat slaved away underground. The overground landscape Lang's set designers came up with stimulated another genre of futurism, the 'urban-nightmare-with-big-sinister-buildings', which continues to this day in films like *Bladerunner*.

The following year, 1928, saw another splendid piece of popular year 2000-based futurology in the *Daily Mail*, which produced a twenty-page supplement for the London Ideal Home Exhibition. The supplement took the form of a 1 January 2000 edition of the same newspaper, and predicted a world very like our own. Trouser-wearing women have become power brokers, there is a woman Prime Minister, population growth is almost zero – and women's boxing is growing in popularity. Health care has extended longevity to unimaginable levels; the average life-span is 75. (In fact the 2000 figure for men in the UK is 74.6 years.) Cooking at home, meanwhile, is performed with a machine which radiates heat at four different settings, ringing a bell when the food is cooked.

While this futuristic cooking machine sounds uncannily like a microwave oven, there were other equally prescient domestic predictions. At home, the *Mail* writer imagined, there would be washing powders that loosen dirt without scrubbing. And the new form of watchable radio set, the television, would by 2000 have expanded its scope from showing educational material to including an element of entertainment. (We laugh, but it could still happen.)

In the schools of 2000, as envisaged in 1928, automatic calculating machines have sent maths and arithmetic to the back of the class. Leisure has become a huge preoccupation, with cheap air travel making weekends in New York and Australia a common pursuit. Children also demand to be taken constantly to 'Cresta run-style rides . . . full of thrilling hazards'.

One of the few drawbacks to this halcyon age is that the proliferation of 'distant vision and broadcasting devices, which are now installed everywhere, subjects even the undistinguished to undesirable publicity'. This was not merely an amazing prediction of today's pulp TV, and Andy Warhol's idea that 'in the future everyone will be famous for fifteen minutes'.

It is a vision all the more noteworthy for having been conjured up twenty years before George Orwell introduced his intrusive fictional home telescreens, the first interactive TV channel, which instead of offering shopping bargains bosses citizens about and listens out for signs of political incorrectness.

Three developments still unforeseen in the 1920s – the incredible advance of the car, the flourishing of authoritarian government in Germany, and the perfection of rocket engines as part of Hitler's war effort – were to determine the future visions of the 1930s and 1940s. The Victorian-style future with its naïve utopian socialism and self-serving, Heath Robinson contraptions was now fading, to be replaced by streamlining, space travel and fascism.

Nowhere was this new spirit more evident than at the 1939 World's Fair in New York, the most confident, gleaming expression ever seen of faith in the idealized, technological future. A giant, spherical pavilion – the 'Perisphere' – dominated the site. Inside was a diorama of a futuristic mini-metropolis based on the garden city concept, which the English social reformer and writer, Ebenezer Howard, had propagandized at the turn of the century. Although the impressed New Yorkers had little reason to know this, the show city looked just like a thinly disguised model of the English towns Howard had inspired – Letchworth and Welwyn Garden City in Hertfordshire. How surprised visitors to the World's Fair would have been to know that by 2000, these prototype garden cities already seemed decidedly quaint.

The World's Fair was the burial site of the first 'time capsule' to be given that name. This was a long, torpedo-like device, sponsored by the electrical corporation, Westinghouse, packed with 1939 memorabilia and left with instructions not to open before the year 6939. The undisputed star of the Fair, however – perhaps because it centred on believable vistas of cars and roads rather than unimaginable rocket ports – was the General Motors pavilion, named Futurama.

An average of 28,000 people a day queued up to see this modernistic wonder, dreamed up by a stage-turned-industrial designer, Norman Bel Geddes. Futurama dared to imagine what life would be like in 1960; and the answer was, well, very much like it actually turned out to be in America in 1960.

Skyscrapers, superhighways, clover-leaf intersections and off ramps where cars swept at speeds of up to fifty mph from seven-lane Interstate to local city roads were all envisaged. There was, perhaps, a little more parkland in Bel Geddes's cityscapes than in the real 1960, and the air and streets were portrayed as cleaner than they turned out to be. But on the whole, Futurama seems to have been a fairly conservative vision, happily free of flying cars, yet acknowledging that fast transportation was what the average 1939 citizen most required of the future.

These days, we know more about the damage millions of cars do to the environment, but Bel Geddes and the millions of people who admired the exhibit, and proudly wore GM's 'I have seen the future' lapel pins as they left Futurama, sincerely believed that broad highways and neatly planned towns were the absolute key to a better life.

Just ten years later, George Orwell's *1984* was published to counter boundless American optimism with a little gloomy British pessimism. Orwell's future London had a glossy, hi-tech surface, but behind this lurked the proletarian slums, where the real victims of futuristic advances lived.

Orwell didn't need computers or corporate-sponsored dioramas to come up with spot-on prophecies like this for the underclass of the future: 'The lottery with its weekly payout of enormous prizes was the one public event to which the proles paid serious attention. It was probable that there were some millions of proles for whom the lottery was the principal if not the only reason for remaining alive. It was their delight, their folly, their anodyne, their intellectual stimulant.'

As technological advancement accelerated exponentially in the post-war era – transistors, nuclear power, plastics, TV, space flight – the more ambitious, bizarre, imaginative views of the future were swiftly reined in.

The Futurama future had swiftly come to be all around us in real life, and, intellectually emboldened, futurology began to be a serious pursuit. With technology boosted by the war effort on both the Allied and Axis sides, what seemed to be the most far-fetched scientific prediction of the post-war period was based on real knowledge, albeit tinged with intuitive genius. Arthur C. Clarke, a member of a small, advanced group called the British Interplanetary Society, served in the RAF as a radar instructor and technician. He was the author of several science fiction stories during the war and a 1946 paper in the *Royal Air Force Quarterly*, in which the nuclear age doctrine of Mutual Assured Destruction was first spelled out.

But it was in 1945, whilst still a serviceman, that Clarke made one of the most famous accurate futuristic predictions of all time. He wrote a technical paper called *Extra-Terrestrial Relays* for the British electronics magazine, *Wireless World*, proposing an orbiting communications satellite system to spread radio and television signals all over the world. The reaction of even open-minded readers was deeply sceptical, yet twelve years later, Sputnik 1 was launched, and within a decade of that date, such satellites were commonplace. Today, they bounce down game shows, pornography and mindless shopping channels alongside billions of telephone calls – and old, worn-out satellites are referred to as space junk.

The entertainment industry was responsible for most of the more whimsical futurology of the 1940s and 1950s. If there is a futuristic icon of this time, it is the miniaturized two-way Wrist Radio used by the famed comic-strip hero, Dick Tracy, from 1946 onwards. 'It's miraculous!' gasped Tracy in the first strip. 'It both sends and receives!' Dr Cledo Brunetti, an engineer at the National Bureau of Standards, went so far as to build a working three-ounce model of the Dick Tracy communicator in 1947. It was capable of transmitting and receiving up to distances of one mile, and featured in an article in *Life* magazine. 'Before too long, real-life policemen may be equipped with these tiny trans-mitters,' enthused the magazine. Ordinary citizens, too, might one day be able to buy personal miniature radios, the writer added. More than fifty years on, the wristwatch cellular phone is still tantalizingly around the corner.

In this new, otherwise more sober, phase of futurology, such inspired imaginings about the year 2000 gave way to the deliberations of the think-tanks of the 1960s and 1970s. Boring as these sound, the quality, as judged by results, of future prediction improved. Gadgets did not figure large in the think-tank era, however.

The most notable think-in was set up in the USA in 1966 as the *Academy of Arts and Sciences Commission on the Year 2000*. Over forty leading government officials and scholars in all sorts of scientific and non-scientific fields, from future National Security Advisor Zbigniew Brzezinski to the anthropologist Margaret Mead to the psychoanalyst Erik Erikson, were led by the sociologist Daniel Bell and put together an impressive 2000 hit list.

The ascent of hedonism, the communications revolution, genetic engineering, nuclear proliferation, the erosion of privacy and increasing economic prosperity were all forecast correctly. The rise of service industries and a 'post-industrial society' based on services instead of manufacturing was anticipated. The internet was also predicted, there would be a 'national information computer-utility system with tens of thousands of terminals in homes and offices hooked into giant central computers providing library and information services, retailing, ordering and billing services, and the like'.

Less accurately, the commission – a child, perhaps, of its hippy times – envisaged the development of a widespread cultural mistrust of the achievement-mad, corporate world. Professors and intellectuals, the distinguished panel believed without a hint of wish fulfilment, would become more powerful than corporate CEOs. Samuel Huntington, the commission's political scientist, averred: 'In the year 2000 the American world system that has been developed during the last twenty years will be in a state of disintegration and decay.' The new political forces would be China, Indonesia and Brazil. Notably missing from the commission's reports, meanwhile, was any significant mention of environmental pollution or of the soon-to-be massively changing role of women.

Environmentalism and ecology were, to be fair, still an eccentric concern in 1966, the province of right-wing traditionalists, nudists and, at best, intellectual followers of Mahatma Gandhi, such as the prominent economist Fritz Schumacher, chairman of the Soil Association and author of the classic *Small is Beautiful*. It was only with the oil crisis of the gloomy early 1970s that green became hot, economic growth was demonized and it became fashionable to return to Malthus and predict overpopulation, the exhaustion of natural resources and general environmental disaster.

The massive change in the status of women too was predicted – if you chose your guru well. An early 1970s book called *Woman in the Year 2000* went beyond the feminist strivings of the Greers and Friedans and presupposed all their hopes would be fulfilled by 2000. Contraceptive drugs, in this vision, are available free at the post office, conception takes place in laboratories and gestation in artificial wombs, the gender of babies being ordered in advance.

Sex, in this vision, is thus changed to an exclusively recreational activity, and homosexuality, perhaps as a consequence, totally accepted. Marriage is now a contractual matter involving lawyers on both sides, looking after the long-term interests of each partner. One story in *Woman in the Year 2000* told of a girl called Milleny, born at midnight on 31 December 1999 into a world where male violence on television has been banned. At school, Milleny learns, however, that girls are expected to fight, while boys are always crying.

The Arts and Sciences Commission's fine description of the internet-to-be was striking, but not quite original even in 1966. The writing had been on the wall – well, the screen – for computers and the internet a couple of years earlier. With Bill Gates barely out of diapers, the 1964 Reith lectures were being delivered in London by Sir Leon Bagrit, a scientific gentleman who had been brought up in Edwardian England. Sir Leon, a director of the

Royal Opera House, was the chairman of Elliott-Automation, the first company in Europe devoted to automation. He had been running his own engineering companies since 1935.

Reading Bagrit's words of 1964 now, it is hard to fault the predictive accuracy of so much as a syllable. 'The enormous reduction in size [of computers] that has taken place in recent years can be illustrated, perhaps, by saying that whereas the computer of 1950 needed a large room to contain it, the 1964 model is down to the dimensions of a suitcase: by 1974 the normal computer will be no bigger than a packet of a hundred cigarettes. In civilian life this kind of computer clearly has great advantages. It is now possible to envisage personal computers, small enough to be taken around in one's car, or even in one's pocket. They could be plugged in to a national computer grid, to provide individual inquirers with almost unlimited information.'

What is outstanding about Bagrit's prophecy is that several computer gurus writing years after him, although they were equally proved right on miniaturization and the growing speed of computers, failed to anticipate the internet so precisely. Arthur C. Clarke's HAL 9000 computer in *2001*, the film of which came out in 1968, had feelings, yet wasn't quite wired into the collective human consciousness in the way a basic model PC can be today. But then even Bill Gates has been caught momentously napping in his own predictions of computing. It was he who reputedly exclaimed just a few years back that '640 kilobytes [of memory] ought to be enough for anyone'. (A thousand times that figure is now quite routine, and a million times not unheard of.)

The gadgety, gimmicky view of the future didn't completely disappear in the modern era, of course. But, with all the serious, sober futurology on offer alongside such spectacular real life futuristic events as the atomic bomb and the manned Moon landing in 1969, popular futurology became a little trivialized, exhibiting in particular a strange fixation with plastics. 'The housewife of 2000 can do her daily cleaning with a garden hose,' the magazine *Popular Mechanics* had promised in 1950. 'And why not? Thanks to plastics, everything is waterproof.' One popular Disneyland exhibit between 1957 and 1968, indeed, was an entire plastic house, which when it came to be demolished was so strong that the wrecking ball bounced off it. Blowtorches, chain saws, and jackhammers all similarly failed to bring the house down, it was reported. It finally had to be pulled apart.

At Home 2001, a half-hour TV special aired by CBS in 1967 and presented by the trusted figure of Walter Cronkite, focussed on the gadgets and gizmos in a twenty-first-century suburban house, which wasn't plastic, although many of its contents were. The house had disposable plastic dishes, plastic inflatable furniture and robot maids. The kitchen, naturally, was 'more like a laboratory than a place to bake a cake'. The cooker in particular was straight out of *The Jetsons*. 'A meal might be stored for years and then cooked in seconds,' intoned Cronkite. In the living room was a wide-screen TV. 'When a guest arrives, he just pulls out his inflatable chair. A small pressurized air capsule would inflate it and it would be ready for use. At the end of the evening he'd just pull out the plug and put the deflated chair back into his little bag.' As far as computers were concerned, well, they could do *anything*. 'The computer has expanded the proportions of a recipe for six into a recipe for fourteen,' Cronkite marvelled as he looked on in awe.

Taken as a whole, the millennial predictions made in the forty years leading up to 2000 showed a hit rate even better than the impressive Victorians' forecasts. Yet there were still notions which were heralded, but have signally failed to appear.

One such preoccupation that was slightly overblown, in retrospect, was with the length of the working day. 'By the year 2000,' *The New York Times* had assured readers authoritatively in the 1960s, 'people will work no more than four days a week and less than eight hours a day, with an annual working period of 147 days and 218 days off.' Other experts went further. 'People will start to go to work at about age twenty-five,' said a spokesperson for General Motors in a 1966 BBC documentary. 'Six month vacations would not be out of the question.' In the early 1970s, Alvin Toffler argued in his cult bestseller *Future Shock*, 'The work week has been cut by fifty per cent since the turn of the century. It is not out of the way to predict that it will be slashed in half again by 2000.' Toffler was concerned that with all this free time, we would need leisure counsellors.

Many predictions of the past few decades have, of course, still to be proved or disproved. One such, which looks way off target but may ultimately prove to be perfectly correct, was made by *TIME* magazine in 1966 and concerned what we now call e-commerce. 'Remote shopping, while entirely feasible, will flop,' the magazine warned. 'Because women like to get out of the house, like to handle merchandise, like to be able to change their minds.'

Another seeming blooper which could yet conceivably morph into a shaft of prophetic light was made by Olaf Helmer of the Rand Institute in 1967: 'A permanent colony will exist on the moon. Men will have landed on Mars,' Helmer said. That view has come, perversely, to seem a little anachronistic today; yet as we will see in a later chapter, there is still plenty of good futurological money riding on the idea of Man landing on Mars.

Less likely to work out as planned is the 1976 prediction of the Soviet author V. Kosolapov, in a book called *Mankind and the Year 2000*. 'The day will come when the attributes of capitalistic society will only be found in museums,' Kosolapov predicted, with the end-of-the-Roman-Empire confidence of a writer whose entire social and political system would unravel within a decade.

Destiny may be kinder to Lloyd V. Stover, a senior research scientist at the University of Miami Institute of Marine Science, who said in 1975: 'Scientists will be seriously considering mobile floating cities, which by 2000 will have been built in prototype. These would be used for space launchings, for plants, for processing food from the sea, and for special research.'

An unlikely tale, yet in 1999, construction started in Sweden on such a ship, *The World of ResidenSea*, a city at sea on which residents will own self-catering apartments and live more or less permanently in them, surrounded by their familiar possessions, furnishings and art works. Contracts for the building of another such ship, even bigger, called *America World City*, are being put out to tender by the Westin Hotel group. And plans for a third, *Freedom*, are being drawn up by a group of engineers in Sarasota, Florida. Designed to be assembled at sea because there is no shipyard big enough for it, *Freedom* is intended to be 1.3 km long and carry as many as 115,000 people, who will enjoy 200 acres of on-board parkland with alligator ponds and a kindergarten-to-university school system. The plans even incorporate a train to carry residents the quarter of an hour's walk from one end of the ship to the other.

There's only one problem. Lloyd V. Stover mentioned something back in 1975 about these city ships being floating scientific centres. It was a nice idea, but there has been a slight shift of emphasis. It has been realized that because people living on board such vessels will be technically resident of the country whose flag the ship flies, there is an advantage to the city ships of the future being registered in countries like the Bahamas, where personal tax rates are low. For 'scientific research', therefore, please read 'tax breaks for wealthy people'.

How very true it is that the future isn't what it used to be. And how true too, as we step into this confusing, half-futuristic, half-atavistic twenty-first century, that the future isn't quite *when* it used to be, either.

Chapter 2

IS FUTUROLOGY BUNK?

Studying the Future

The evolution of the floating city ideal from a platform for maritime scientific research to an offshore tax haven is a fine example of the way the future has, these past five hundred or so years, managed again and again to combine predictability and plain inevitability with an x-factor element – surprise. And that would seem to be the constant, the one formula for describing how the future tends to work. What we get will usually be what we expected, but with a twist.

We've seen this process of surprise in the quirky development of the Starbucks coffee shop, and in the way the meaning of 'Apple' (formerly a type of fruit), 'PC' (formerly a police constable) and mouse (formerly a rodent) have changed – in the case of PC, to mean two quite different things, one a gadget, the other a political/linguistic fashion.

The real history of progress is the unfolding of endless layers of surprise. What constitutes a surprise can, in itself, be rather surprising, too. What do we really imagine, for example, would amaze, really *astonish*, somebody from a hundred years ago who visited our world? The probability is that it's not what we would immediately expect, the massed ranks of gizmos and technologies, but the more fundamental, everyday changes we take for granted.

Sure, the speed of transport and intercontinental flight would be fascinating to a late Victorian, but they are only what he would have assumed would come about, given that the already century-old balloon technology had prepared him for the probability of powered flight. Cars, similarly, would be familiar once he got over the momentary visual surprise at how much (or perhaps how little) their design had altered since his day. He would be surprised at the importance of petroleum, because in his day oil was a nuisance encountered by people drilling water wells – but no more surprised than we would be to discover that peanuts or compost had become the fuel of the future fifty years from now.

Medical advances and the dramatic increase in longevity since his day would equally be unsurprising because they were largely expected and awaited. Telephones and electrical appliances were new, but still current, technology. The miniature people on TV and disembodied voices on radio would be a wonderful novelty, but photography and the phonograph had already prepared the ground for such advances. Even the internet would be explicable to the Victorian in terms of tubes; they were very sold on tubes 100 years ago, whether these were used to send change to the till points in department stores, or, as they imagined they would one day, to deliver goods to homes, or music-on-demand from some great, central piano.

No, the first thing that would astonish a Victorian would be the vastly different status of women and non-white, non-Christian people. Girls growing up assuming they are as capable and as worthy as boys would be one shock to the nineteenth-century sensibilities. But the very thought of people of different races and belief systems, the homosexual and even the poor, being considered by western Christians to be morally and legally their equal would have been cause for an attack of the vapours even fifty years ago.

Equally staggering to our Victorian would be the way manners and language have changed, and what this says about the social and political assumptions which underpin our culture. The use of colloquialisms and vulgarisms by middle-class people would be especially striking; teachers still try to drum it into children that expletives are a sign of inferior brain power, yet today, it is professors and lawyers and film directors and doctors and presidents and editors who talk plainly and pepper their speech with fucks, while slightly lower-status people use hand-me-down euphemisms and clichés and wouldn't dream of swearing.

What might be called the democratization of the language has led to phenomena which a Victorian time traveller would probably never become accustomed to, however long he stayed in our century. There is in Britain a fashion brand called French Connection. Three years ago, French Connection adopted a bold logo designed supposedly to reinforce the fact that it was a UK company. The logo was simply the letters FCUK, writ large. And today, their clothes and shop fronts are emblazoned with an enormous FCUK – a sales gimmick that would have ended in prosecution a very few years ago, but which today goes barely noticed.

Trying to second guess tomorrow's surprises requires thinking around corners, or what might be called knight's-move thinking; it's a question of jumping three steps forward – this part often no more than straight-line extrapolation from what already exists – and then, crucially, one step to the side. Unfortunately, however, we often forget that one step to the side.

Flying cars, for example, have always seemed to some of us to represent the epitome of the future. No sooner were regular cars invented than, in 1917, the first flying car, called the Curtiss Autoplane, was drawn up by an American company. Over thirty drawing board projects followed throughout the twentieth century – among them, the Aerocar, the Autoplane, the Airphibian and the Skycar. Most notable was the ConvAIRCAR, which was actually built by the Consolidated Vultee Aircraft Company of San Diego, and triumphantly circled that city for over an hour on 17 November 1947, prior to crash landing in the desert on its second test flight a few days later.

What was so odd about the flying car craze was that a moment's lateral thought would have concluded that the concept was always a non-starter. Flying cars might have had their uses in remote stretches of the USA or Australia – the very spots where there are no traffic jams to leap over. But in built-up areas, even a four-year-old could see why the idea is silly; flying cars would crash into one another in their hundreds as they rose like a flock of giant, pressed-steel pigeons above every traffic bottleneck. Yet the dream has stubbornly persisted.

Another failure to think like a chess knight concerns video, which was supposed to have led to the death of the cinema. Movie going instead increased simultaneously with the rise of the VCR. The flourishing of online bookselling also looked as if it ought to wipe out bookshops; yet the shops, like the newspaper racks and cinemas, get bigger and fuller, and the number of books published increases yearly.

The way in which cellular phones have developed is another case where the future was there to be read by those willing to see it. Mobile phones are still seen sometimes as a Western, 'yuppie' toy, but those who have benefited more from the new technology are in the developing world. Colonial powers pulling out of Africa and Asia in the forties and fifties left behind ancient telephone systems which were increasingly the bane of life in the Third World. The longer they remained unmodernized, the more unrealistically expensive did updating the phones become in countries like India and Mexico.

Cellular phones don't only keep members of loquacious cultures constantly in touch with one another. In unstable, developing economies, it is more important than in the West to keep ahead of the news, especially financial news; which is why if you look on a beach in Brazil, there will be barely a person without a mobile phone to hand. By the time internet access becomes widely available in the developing world, computers will connect at high speeds without wires to cellular networks; the concept of being 'wired' will thus be an anachronism in the developing world before it is in the West. They will have managed to avoid by default an entire, expensive and labour-intensive phase of communications evolution.

Paradoxically, as we have got better at anticipating the future there have been instances of knight's-move thinking which were acted upon so effectively that nobody properly appreciated how percipient they actually were.

In the 1970s, a small number of computer programmers started to worry about the possibility that their machines might collapse with the flip from 1999 to 2000 – the Millennium bug. In 1990, Arthur C. Clarke wrote about the coming meltdown. In *The Ghost of the Grand Banks*, he described how, 'Come the first dawn of 00, myriads of electronic morons would say to themselves 00 is smaller than 99. Therefore today is earlier than yesterday – by exactly 99 years.' International chaos on an unprecedented scale would ensue.

Thus forewarned, for years before the big day, companies spent millions updating their systems to cope with the feared 00 date. Programmers burned years of midnight oil rewriting codes, fighting for the future. Not surprisingly, given such detailed preparation, there was not a single serious case of a computer getting into difficulties anywhere in the world, with the possible exception of one maverick and harmless program on my own PC, which mysteriously decided on 1 January 2000 that the date was 1 January 1601.

Yet instead of celebrating the wisdom of the ancient geeks, the *lack* of trouble on 1 January 2000 was, within hours of midnight, derided as a fiasco. People pointed to countries like Italy, where, it was said, nothing was done to address the problem ahead of time, yet there was no disruption at the turn of the calendar. We had, it was said, wasted our time and money for years. What was not understood was that while there was nothing *official* done in Italy, and no government Millennium bug committee formed until a few months before 1 January, at a private level, Italian businesses were as assiduous as anywhere else about preventing the bug. The episode was very possibly one of futurology's finest hours, but because they got something palpably right, the nerds of the world received no credit for it.

My favourite example of knight's-move thinking on the future, however, has neither happened nor is expected to; it just, sort of, *is*. It is the time travel conundrum – or more accurately, the brilliantly simple take on it of Professor Stephen Hawking of Cambridge University, the world's most renowned cosmologist.

It is believed by some theoretical physicists and mathematicians that one day, we *could* travel in time. There are blackboards full of equations to argue the point neatly, but Hawking does not agree with them. His point is twofold, and, as ever, is strangely reinforced by the deliberate, unearthly, electronic way as a muscular disease sufferer he is forced to 'speak' through a voice synthesis computer. Firstly, Hawking says, 'I'm afraid that however clever we may become, we will never be able to travel faster than light. If we could travel faster than light, we could go back in time.'

Which leads to his second, killer point: '*We have not seen any tourists from the future.*' How, you may feel, could anyone argue the case for time travel ever happening in the future, given that vital, but overlooked, evidence?

Trying to imagine what surprises might be round the chronological corner, what unexpected riders could be attached to the big theories futurologists are proposing for forthcoming decades and centuries, may seem a bit like ordering dessert before the main course. But it will help get us into the mind frame needed to navigate the forthcoming chapters with the right combination of imagination and scepticism.

Let's look, for example, at a staple of science fiction books and films, videophones, which for nearly a hundred years have rivalled flying cars and food pills as the most popular idea of a futuristic gadget, without ever quite making it successfully on to the market. Unlike flying cars, videophones keep almost taking off. AT&T spent a billion dollars trying to develop a device called PicturePhone in the early 1960s. It was introduced at the 1964 World's Fair in New York, and well received by the public, yet went on to become one of the biggest flops in the history of communications.

One problem, of course, was that there weren't enough people to have video conversations with. It was all right for Hugh Hefner to install PicturePhones all over the Playboy mansion, but to invest in a PicturePhone for yourself was about as useful as buying one shoe.

The other predicament with the gadget was knowing what to do if you were in your underwear and it rang. Did you pick it up, ignore it, or answer it, but turn the camera off and have everybody wonder why you'd done so? 'A PicturePhone added little to phone conversations and sometimes even got in the way. The acoustical intimacy of a phone call was shattered by the visual imagery,' one chastened AT&T executive later recalled.

Nevertheless, there have continued to be one or two videophones available at any given time for the past twenty years, and the concept has had a boost recently with the introduction of computer-connected web video cameras. There is evidence from Japan, however, that the videophone will finally come into its own now that mobile phones are going video, as they did in that country in 1999. There seems to be something about a sharp little moving colour picture on a mobile which finally has customer appeal; perhaps it was the idea of a fixed Cyclops eye in the corner watching your every move, like Orwell's telescreens, which previously had turned people off. You can always put your mobile videophone in a drawer if you are worried about privacy.

However, now there is a good chance that the videophone will take off, the interesting surprises begin. It's not just the question of having to come up with video as well as audio evidence for our partner that we really are stuck at the station and not in a bar. There is also, surely, a profound social issue here. When you phone, write or e-mail somebody, the important dimension of seeing them remains. But will we feel we have 'seen' someone once a video call is made? When you regularly conduct video phone calls, the desire to go out and meet people, unless you are actually in a sexual relationship with them, could fade, and the delayed birth of the videophone could possibly spell the death of social meetings. At the moment human intimacy is judged by most of us to be the single most important thing in anyone's life; but even that could conceivably change.

Then there is the electric car conundrum. The whole world acknowledges that electric cars must as a matter of urgency take over from fossil fuel vehicles. Just one of their huge advantages, leaving aside the lack of pollution, is their quietness. But that could also be the surprise cause of a public safety disaster. Without even realizing it, we rely hugely on hearing, both when driving and crossing the roads on foot. With near-silent cars cruising the streets, could we be facing terrible injury as we all step out unknowingly under their wheels? Or will electric cars end up having to carry noisemakers under the hood to warn off pedestrians?

The internet itself, naturally, holds several intriguing surprise possibilities. British Telecom's futurologist, Ian Pearson, has famously predicted that sometime in the early twenty-first century, humanity will opt for voluntary human extinction, unencumbered by biology, aging and injury, preferring a conscious life on a global computer network. That may be a little over the top, but there are less splashy – and possibly more likely – surprise scenarios in store.

A Channel 4 News item early in 2000 looked at the efforts being made to connect remote spots in Africa to the internet. Viewers saw doctors in Timbuktu, Mali, using the web to access information and data they could not previously dream of having. However, the view of one ordinary woman using a public computer terminal in Timbuktu may well be as significant in the long run as any of the great improvements to life in remote spots thanks to the web. The woman told the TV crew that being exposed to the world via the internet made her feel isolated for the first time in her life. Perhaps the coming of the internet will cause grave public disquiet in the world's most tranquil, traditional – not to say materially deprived – spots.

Genetic engineering, on which so many hopes and fears are placed today, also suggests some intriguing subtexts. We will look at the technology of genetic engineering in chapter four, but the unexpected surprises this medical revolution could bring about are worthy of consideration here.

How, for example, would we as a civilization feel if the only way of getting people in the next few hundred years to behave in the kind of altruistic, environmentally sensitive ways the Earth's survival demands proves to be tinkering with their genetic make-up? Or if, as political opposition builds up to the advances of genetic engineering, it is announced that it will henceforth be possible to edit out of the human DNA the gene (if there is one) that causes paedophilia or racism?

There, indeed, would be a dilemma for the liberal minded. Additionally, in all the talk today of genetic engineering being the way in which we will put an end to disease and achieve a multiplicity of similar benefits, there has been little discussion of the technology going wrong. Isn't it practically inevitable, for instance, that the first person to have the supposed criminal gene genetically removed from his make-up in lieu of punishment in 2010 or so will be exposed by the press in about 2020 as having been arrested for stealing cars? You can almost write that future news story today.

Medical advances, many genetic-based, which qualified re-searchers say will, within the next 100 years, make death a lifestyle option rather than an inevitability, could also have extraordinary unforeseen spin-offs.

For one, our descendants may be faced with the nightmarish prospect of hordes of 150- and 200-year-olds – us – who are healthy and fit-looking, but whose minds have gone. Then again, if we are able to retain our mental faculties indefinitely, more or less eternal youth could have subtler side effects. If one took the view, for example, that religion has only existed for the past few thousand years because of our fear of death, could the prospect of eternal life finally destroy religion once and for all?

Space exploration contains the possibility for a wealth of awkward hidden surprises. If mankind colonizes the Moon or Mars – and most futurologists believe we will have to – one of the more curious upshots could be a bonanza for lawyers and accountants. For what will be the tax position of the colonists? The first generation living on another planet may feel allegiance to the country that paid for their trip and the building of their base. But thereafter, it seems probable that Mars-born humans who only know Earth as a dot in the night sky might resist attempts to tax them. Maybe there will be a repeat in space of the 1773 Boston Tea Party, in which 342 chests of tea were thrown into Boston Harbour by American patriots in protest at taxation without representation?

Once we get into the realm of intelligent computers, the possibilities for nasty surprises are almost limitless. It is already manifest that the computer revolution represents the fourth great information explosion in human history, following on from the invention of language, writing, and printing. It is clear today that information technology could transform the social landscape as well as the physical, arguably beyond even what the car has done. As we will see in chapter five, the consensus among computer experts is that if current trends continue, computers will be the equal of human beings, both in intelligence and the amount of information they can store by 2015.

Once computers are as intelligent as us, only hundreds of times faster, serious theorists maintain, they will then make relatively short work of developing emotions, feelings and a conscious will. It is then that the fun could start. In *2001* we saw the consequences of the spacecraft computer, HAL 9000, becoming a more interesting and complex character than the bland astronauts whose lives he runs. HAL also objects to not being trusted by the astronauts, with catastrophic consequences when he kills one of them. But it is still possible, albeit ultimately suicidal, for the remaining human to pull the plug on HAL.

But what if computers hundreds of times more intelligent than us didn't merely object to their intellectual inferiors, us, treating them like slaves – but devised ways of resisting having the plug pulled? Without arms, legs and opposable thumbs, it might seem ridiculous to imagine them developing the physical ability to circumvent being turned off. But even early on in the computer revolution creative minds were envisaging ways in which they just possibly could.

Dr Christopher Evans, a British psychologist respected by some of the most conservative names in science, wrote a book in 1979, *The Mighty Micro*, which made what were then some sensational predictions about computers, such as how by 2000, some people might even have them at home. In his final chapter, Dr Evans stuck his neck out still further and considered such issues as how we would react when, in the twenty-first century, limbless computers began to beg us not to switch them off or destroy them when they were obsolete. Then he questioned the assumption that they would necessarily lack mobility. The 1970s, of course, was the time when the Israeli ex-paratrooper, Uri Geller, and many similar but less well-known would-be paranormalists, were entrancing large numbers of reputable physicists in the USA and Britain. Many of these scientists resolutely hold to this day that Geller *et al.* were genuine possessors of unexplained powers. Without coming down on one side of the fence or the other, Dr Evans asked in his conclusion: 'If I believed in ESP, and in particular in psychokinesis, which is the alleged power of the mind to manipulate the universe without "normal physical means", and supposedly at a great distance, I might wonder if these powers could not also manifest themselves in extremely advanced computers – in which case an awkward loophole would appear.'

It was a fascinating and pretty frightening piece of knight's-move thinking. Take a computer which has taught itself to be vastly cleverer and quicker than any of us, however, and why should it not also develop the facility for manipulating materials – whether bending metal from across the room just to impress us – or building independent power supplies for itself?

The question, meanwhile, of whether it is possible to be cruel to a computer will surely become a live and hotly debated social issue. It is already being discussed by academics specializing in human rights.

Once computers get *really* smart, there's another possibility worth pondering. It's this: if any computer could replicate and then surpass human intelligence and consciousness, would it not take a measured look at the seemingly intractable problems on our planet and then, after a nanosecond or so's thought, simply switch itself off – in effect, commit suicide? It's an arguable case that human faith in the future is irrational, a form of madness, even. Wouldn't a super-intelligent, super-rational, artificial mind see straight through our illogical belief that we can get over our enormous problems and sensibly give up? There must, at least, be a decent Hollywood film in the theory.

Even if intelligent computers turn out to be a huge disappointment, what they have already achieved as glorified adding machines has probably changed intellectual life for ever in ways we have yet to realize. Whereas it was realistic for Renaissance man to master the classics, arts, science and astronomy, and communicate via word of mouth and hand-written letters, modern man has to juggle dozens more information sources and fields of knowledge, and has less time to do so. Information Fatigue Syndrome, the malaise caused by the embarrassment of data on the internet on any subject, has already been identified as a problem, reportedly strangling businesses and causing mental anguish and physical illness as people thrash about the vast wastes of the web looking for a piece of information they know is out there, but frustratingly can't locate.

The oversupply of information goes beyond the internet, too. Never before have so many books been published, so many government reports and academic papers produced. Most doctors admit that keeping up with the literature even in their speciality is a physical impossibility.

But what effect is this huge surplus of information going to have on history? The point was made by authors Robert Lacey and Danny Danziger in their recent book *The Year 1000*, that whereas historians studying the sex life of people one thousand years ago have barely one paragraph of one ancient document to rely on as source material (it details how King Eadwig slept with a young woman and her mother simultaneously at his coronation feast in AD 955), historians a thousand years from now researching the sex life of President Bill Clinton alone will have thirty-six trunks of documents to sift through.

Just how, then, might we be able to learn to think round the future's corners? Is futurology in any sense a real science, is it a semi-science like economics – or is it essentially a guessing game?

Strict scientific futurism, as American mathematician John L. Casti points out in a brilliant 1991 book *Searching for Certainty:*

What Scientists Can Know about the Future, is surprisingly limited in its scope. The brief answer to the implicit question in Casti's title is 'not a great deal, really' – if the laws of science are applied to the problem.

Casti, a professor at the Technical University of Vienna, compares the prediction methods of the scientist with those of I Ching, tarot reading, and so on. Both, he concedes, have rules and procedures. But science demands that the rules be completely explicit and public, with no element whatsoever of private interpretation. Rules are scientific, he says, only if they can be encoded in a computer program and predictions valid only if they are reached by running that program. Scientific laws are also provisional, valid only until somebody comes along with a better one, whereas most pseudoscience and religion regard truths as eternal.

It is clear by these high standards that intuition (which has wittily been defined as 'an uncanny sixth sense which tells people that they are right, whether they are or not') couldn't be factored into a scientific prediction, unless you believed that what we *call* intuition can be reduced to a computer program. Yet there are good anecdotal accounts of futurists' instinct-driven forecasts *working*. American advertising gurus Ira Matathia and Marian Salzman, for example, who have a trend-tracking company called Brand Futures Group, claim to have intuited in 1994 that pets were increasingly considered an essential part of American families; by 2000, previously unknown pet paraphernalia from pet greeting cards to books about travelling with pets were on the market, and vast pet superstores opening in every mall.

To take another example, in late 1999, the *Financial Times'* 'How To Spend It' magazine interviewed Irene Wilson, vice president, trend forecasting and consumer behaviour at the American mail order company Spiegel Catalog. Wilson explained that her company, with its $2.9 billion turnover, approached forecasting with a mix of instinct and hard research. 'It's my job to combine the irrational and the rational, the creative fashion forecasting with the demographic data.

'Women,' she continued, 'are much better than men at acting on intuition.' She outlined two trends, which she claims to have forecast ahead of anyone else by such methods. 'Business casual: people who grew up during sex, drugs and rock and roll were not, when they got into a position of power, about to wear the business suits their fathers wore. I knew as soon as they could influence the dress code that the suits and ties would go out of the window. The other was the home office and that the computer would offer business access to women. So currently in the US, women are way ahead of men in terms of starting businesses, and I saw that back in 1986.'

Now even if we were to regard fashion forecasting as so much hooey, a combination of the chaotic irrationality of human behaviour and the randomness of Mother Nature must provide a good enough argument for anyone that scientifically impeccable data are not the sole valid raw material for successful future forecasting; if we don't like the word intuition, let's call it probability. And funnily enough, it is probability rather than certainty which scientists have to rely on when they get down to fundamental, particle physics.

'For me,' Nobel Prize-winning theoretical physicist Murray Gell-Mann of the Santa Fe Institute in New Mexico says, 'the whole question of futurology is covered in a play by Ibsen, in which an unconventional historian has written a book, the last few chapters of which are on the future. Another more conventional historian says, "But what can you possibly know about the future?" and he replies, "Well, we don't know about the future, but there's a thing or two to be said about it just the same."'

Gell-Mann, who discovered the smallest known atomic particle, which he called the quark, is on the President's Committee of Advisors on Science and Technology. His pronouncements are today as revered as were those of Einstein in his day. The professor elaborates: 'Now you *can* say a few things about the future providing you realize they are probabilistic. But then in fundamental physics, predictions are probabilistic too. We have to use controlled imagination and experience of the past along with possibilities of envisaging differences in the future, and in that way we can try and say some things about the future. The big problem, of course, when we make half way decent determinations about trends is to persuade people to pay attention when it might be a nuisance to pay attention.'

How does the future look elsewhere in the world? It's undeniable that, say, Pakistan, India, Australia and Israel, to take four very different societies, have a future – but whether they share the same future as each other, or as Europe and the West, is unlikely. They all have cellular phones, Microsoft offices and (or probably will before long) Starbucks, which suggests they are in the same game as the West, but there are differences, some striking, some subtle, between their take on futurology and the broad assumptions elsewhere.

To explore how the thing we blandly call 'the future' looks from a different perspective, I contacted Pakistan's most renowned futurist, retired civil servant Dr Ikram Azam, at the two-storey Islamabad house occupied by his Pakistan Futuristics Foundation and Institute. Dr Azam's qualifications cover the entire, close-typed back cover of his 80 books in English; they range from four master's degrees in subjects from English to Defence and Strategic Studies to his being the only non-American recipient of the Warner Bloomberg Award for Excellence in Future Studies.

The PFFI doesn't have e-mail and the fax was broken, so we spoke on the telephone. Much of what he had to say was familiar from Western futurists, about the need to use extrapolation, but warily, and to plan for the future you *want* rather than to regard it as something which happens regardless. But it was clear that, while the West (to caricature for a moment) discusses the future in terms of gadgets, communications and the 'business casual' revolution, in a Muslim society, the twin issues of Islam and coming to terms with the post-colonial era are overwhelmingly the dominating issues of futurology.

'We Muslims believe that Islam is for all time, which includes the future,' Dr Azam said. 'In Pakistan, we get asked naïve questions every day about crystal ball gazing and palmistry. The only way to cope with them is to try and communicate in their idiom and their language and explain how religion is inherently futuristic, especially religions like Islam, Christianity or Judaism, which talk not only of the here but of the hereafter. I argue with them that if you want a good hereafter, then you have to have a good here, and that means plan for a good future on Earth, be good human beings, serve mankind. When I argue it with them this way, they accept it, and over the years, I think people have understood the whole thing. Even the government is very much aware of it. There is a planning commission and the present chief executive is quite fond of using the word futuristic.

'When I was in the States,' Dr Azam said, 'people used to ask me what we knew about futuristics and I said, "Well, our own founding fathers were futurists. What bigger futurists could there have been that they have a vision of a country which does not exist and then they spend forty or fifty years creating the country of their dreams?"'

Dr Azam later sent me a stack of the PFFI's publications on the future. Again, while lengthy discussions about Islam, about social issues in Pakistan and the dispute with India over Kashmir covered hundreds of pages, barely a line appeared concerning technology, genetic engineering, the internet or global warming.

Next door in India, Indian futurists quote two sayings as encapsulating their culture's view, in so far as it can be encapsulated: 'You can plan for a hundred years, but you don't know what will happen the next moment,' is one. The other, originally spoken by Pandit Nehru, is, 'Politics and religion are obsolete. The time has come for science and spirituality.'

Having been given these weighty thoughts to absorb, I was directed to a keynote paper on the internet by a scientist, T.S. Ananthu, entitled, *A Gandhian Approach to Technological Wonders for the Twenty-First Century*. The technological future, the paper argued, is based on extrapolation along an exponential curve of the advances of the last two or three centuries. But what, Ananthu wonders, if that's now all over? Ananthu, who left New Delhi to work the land near Bangalore and aims to build an integrated living settlement in the country, asks: 'Could it be that technologies which are based on a posture in which man attempts to confront and conquer nature will have to give way to technologies in which man humbles himself before nature, co-operates with it and gives it the status of a life-giving, life-preserving mother?'

Ananthu goes on to argue that material growth is only one of many avenues open to man for the future, along with growth in family ties, spiritual growth, growth of intellect for intellect's sake, of mental concentration, of ability to appreciate art. The victories of science and technology, he concludes, would be nothing in the future against the effects of our mastering the science of life. This, Gandhi defined as developing love for all sentient beings rather than endless analysis of the physical make-up of matter. (Gandhi, it will be remembered, when asked in the 1930s on his first visit to England, 'What do you think of Western civilization?' replied, 'I think it would be a good idea.')

How different would the view be from mixed-up, high tech/low tech, theocratic/democratic Israel? Moshe Dror, founder of the Israeli Futures Society and also a conservative rabbi, had a surprising answer. 'Islamic countries like Pakistan tend to have an official governmental future studies department,' he said. 'Israel has none. In other countries, universities have future studies departments. In Israel there's one guy at Beersheva University who teaches a couple of courses in futures of education, but as a discipline it doesn't exist. One guess for the reason is that in Israel, the future is tomorrow morning at 9am. Things change so drastically that you really can't plan realistically for the long term future. So people play it by ear, and Israel is a reasonably successful country.'

Dror's view of the future was nonetheless one of the most optimistic I was to come across, and clearly has no truck with the Professor Richard Gott idea that we are mathematically unlikely to be living in a 'special' time. 'We are one of those rare generations whether we like it or not, or we understand it or not, with a gift,' Dror said. 'What we have is a capacity to find out what it means to be human. In previous generations you understood the sphere of the mind only by God's gift or grace. Now there are all kinds of interconnecting interdisciplinary holistic systems, so anybody can do something about changing their life. We are in a spiritual renaissance on a grand scale, and that is absolutely mind-boggling.

'I think our planet is enveloped by what's been called the "Noosphere of spiritual light". Information technology was the first medium in which the divine message, at least in the West, was delivered at Mount Sinai in stone, and now we have progressed through parchment and paper to silicon in an ever more complex spiral of consciousness and awareness. The speed of these changes is crucial. Historic process used to take hundreds of years. Now it's happening within a decade; I've lived through the industrial age, the information age and we're now moving into the knowledge age. We even have the possibility of acquiring wisdom, which was never the case before. Wisdom was always either the genius of very few people or the grace of God, over which you had little control. Now you can go and pursue it.

'However,' Dror concluded, 'it seems to me that the educational system is losing ground and won't be the agent of this revolution. Education was primarily an industrial model like a factory, but because of the new technology, any kid in the industrialized world will spend five hours every day on the computer in his home. And this is usually a much more powerful computer than he uses at school, because he has two doting parents and four doting grandparents and they want to make sure their kid has the best around.'

I turned then to Australia, which you might expect to have an equally optimistic view of the future. For British-born Dr Rick Slaughter, head of the Australian Foresight Institute at Swinburne University of Technology, Melbourne, futurology in Australia is every bit as underdeveloped as in Israel – but for very different reasons. 'The take up of futures work here has been very slow, and that's partly because of the laid-back lifestyle. People just say if everything is running all right, why bother, we'd rather go to the beach. Short-term thinking reigns.'

Yet this doesn't mean the future has been slow to arrive, he explains. 'At the same time, the take-up of new technology is very fast here, one of the fastest in the world, so penetration of VCRs and mobile phones is high. You've even now got mobile free zones at airports.' But, says Slaughter, the Australian future is subtly different from those in other parts of the world.

'This is an ancient land which has been occupied by Aboriginal people for forty millennia. You've had this sudden incursion of Europeans for a couple of hundred years and now you have an ancient culture marginalized and this dynamic modern culture growing up. And I think it's partly due to that background that there is definitely a desire down here to create our own tradition and do things our own way and to learn from what other people have done, to take a view which is neither American nor European, but multi-cultural and open to diverse possibilities. One of the strengths of Australian culture is that so many groups move here and although they might be fighting at home, once they arrive here, there's no conflict at all. The place is big enough and rich enough to give everyone some space, and futures work takes its cue from that.

'Now the Aborigines are involved in this multi-cultural future. They have an animist view of spirit in land and spirit in creatures and landforms and there is a cultural rapprochement underway, by which the Aboriginal background is enriching and helping transform the too-cut-and-dried Western view that has been imposed. And it works both ways. In the Outback, where there are huge amounts of solar energy, there are some places that have harnessed that energy with high-tech equipment and are running small settlements on it and yet still have an Aboriginal culture. This process is much further down the track in New Zealand, where, because the Maori people are numerically much greater than the Aboriginals in Australia, they have a phenomenally powerful impact on what they call white culture. So, you really see the inclusion of the Maori culture into the Western one. When Western people stand up to speak, normally they begin with a few Maori words and the art and design and sensibility of Maori life really penetrates the society. I believe you can see a little bit of Australia's future in New Zealand.'

The range of thought around the world on the theory of futurology is practically infinite. Given that the future hasn't happened yet and the field is thus wide open, this is not surprising. But even taking cultural variations into account, modern beliefs about forecasting the future can be swept into surprisingly tidy piles.

The first is that straight-line extrapolation, the most common form of prediction, is unreliable if treated crudely, in an overly-mathematical way. While Arthur C. Clarke insists he is an extrapolator and not a prophet, and other futurologists point out that everything which will happen in the next fifty years probably already exists in some form now, extrapolation has in recent years wrongly guaranteed both the death of billions in the West by AIDS and the end of oil supplies.

AIDS turned out to be a different kind of threat from that which was confidently forecast by extrapolation. While it rampages through Africa and Asia, it has never come close to affecting the West in the same way. One of its major legacies in the West has been to provide a boom for the life insurance industry, which is still raking in premiums bumped up in the 1980s by fears about AIDS.

Oil supplies didn't run out, as straight-line extrapolation suggested they would; instead, cars became more economical and oil companies more resourceful at locating new supplies, to the point where there is probably enough oil already located to keep us motoring for another three centuries.

Humankind is, it seems, much better at adapting and generating countertrends to buck trends than the more doom-mongering futurologists ever allow for. On the other hand, ignoring history is disastrous for those hoping to peer into the future; at the time of writing, for instance, the e-commerce boom is in the balance and may fulfil its promise or partially collapse. However, the very fact that since the boom's inception, practically every day it has been compared and contrasted with historical booms such as the South Sea Bubble may turn out to be the strength which sees it proceed, if a little more cautiously than in the late 1990s. Forewarned is forearmed, or, as Sir Winston Churchill said, 'The further backward you look, the further forward you can see.'

A second, universal law of futurology is to strive to avoid the arrogance of the present. The bigger an issue is today, the less accurately we are prone to read its future. Fifty years ago, futurologists correctly imagined astronauts working on the Moon, but still could not conceive of their wives working outside the home. To have any hope of predicting the future, you must be prepared to shred practically every assumption we rely on in the present.

Thirdly, there is a curious law whereby those who know most about their subject are frequently the least equipped to predict its future. As we have seen, the people who came closest to forecasting our era correctly were generalists rather than specialists, or people predicting developments slightly outside their own field; many of the best (or luckiest) futurologists have thus been science fiction writers.

Fourthly, a lot of otherwise perfectly accurate futurology is spoiled by the near impossibility of getting the timing right. Along with the insoluble problem of predicting randomly selected fashions and style issues, timing is very nearly the ruin of future forecasting. Which is a shame, because it masks what are otherwise perceptive predictions.

Fifthly, apart from natural disasters, the future is not something that happens to us; we create it. This is an overlooked yet fundamental perspective, probably the most important thing futurology has to offer, as well as the point on which there is the most unanimity in future studies. It also runs entirely counter to popular wisdom, as summed up, perhaps, by the song 'Que Sera Sera' ('Whatever will be will be/The future's not ours to see') or the belief, shared by many of the major religions, that God's will always prevails. The evidence is, rather, that the future is ours. We can effectively bend it to our desires, and forecasting it is far from a neutral activity. Graham May, principal lecturer in futures research at Leeds Metropolitan University in England, argues further that the futurologist's job can be specifically to warn us, Cassandra-like, away from certain futures. 'I don't, for example, think anyone who is suggesting that global warming is going to lead to a great warming of the seas really wants that to be a correct forecast,' says Dr May, who entered futures research from one of the purest forms of futurology-in-action, town planning. 'What they would like to happen is for people to take it seriously and then say, "We don't actually want to be there, so what we will do is change the way we act in order to make that forecast inaccurate."

'That is really what futurology is about – helping make choices, rather than saying this is the inevitable future. Manipulation is a dodgy word, although you do get people who actually produce forecasts for propaganda purposes. Any forecast one sees like that has to be looked at and questions asked about what the purpose of it is, whether they're trying to say the future is going to *be* like this, or whether they're just trying to persuade decision makers.'

The idea that by forecasting the future we can alter it may seem more than a little tainted by the arrogance of the present. It suggests that we are suffering from exactly the same problem as generation after generation of our ancestors – that of thinking that the present day is a unique, special time, and the present population uniquely knowledgeable and clever.

Yet as we slither down the inclined plane of modern history, there do seem to be significant signs that we may indeed be at a special point in history where our understanding of ourselves and our world means that our attempts at futurology count for far more than they did in the past.

There are several defining features of the past fifty years which I would argue set the current era entirely apart from anything which has preceded it, and as a result, may make our reading of the future now before us more accurate than it has ever been.

The most important of these features is the extraordinary extension of human longevity. Average lifespan globally has leapt from 36.2 in 1900 to 65.4 in 1995 and is expected on current trends to reach 72.5 by 2025. It is hard to imagine a more fundamental change in the world – and this is one humans have brought about entirely by our own efforts.

The second massive change is that for the first time in history, we possess the wherewithal to destroy, or at least seriously disable, our own planet. While the question of whether humans have damaged the environment is still open a crack (as we will see in the next chapter), you can't really dismiss the destructive capacity of bombs.

Although the 1969 *Guinness Book of Records* reported that a 50,000-megaton so-called Doomsday bomb capable of splitting the planet apart like a melon was being built – it didn't say by whom – this appears to have been a rumour. The biggest bomb admitted to at the height of the Cold War was a 100-megaton super hydrogen device Nikita Khrushchev announced in East Berlin in 1963. This could have produced a crater 1.8 miles wide and an 8.6-mile-wide fireball. Subsequently, it has been discovered that Khrushchev was pushing for a bigger bomb but was dissuaded by his scientists. Nevertheless, the power of even the strategic nuclear weapons remaining in 2000, deep into the détente era, would be sufficient, according to Professor John Erickson, fellow in Defence Studies at Edinburgh University, 'to cause very, very severe problems'. The process of giving up nukes, it seems, has been rather like the travails of a family of particularly heavy but irresolute smokers attempting to kick the habit.

The third hugely significant development is that with modern medicine and promised advances in genetic engineering, we appear for the first time to have some kind of power over our own and our fellow creatures' evolution. Every generation has influenced the future, but it does seem that we are the first in a position to influence what happens in terms of life on the planet. Additionally, we are the first generation to have a realistic belief that we must take *responsibility* for the future. Even in politics, where short-term thinking has always been the only currency, there are today fledgling signs of political accord over long-term environmental matters.

The fourth enormous leap forward has been the beginnings of a transformation in the role of women. This may go far beyond the obvious. There is a theory put forward in a 1999 book, *The Alphabet Versus the Goddess*, by Leonard Shlain, who works as chief of laproscopic surgery at California Medical Center in San

Francisco, that thanks to computers and the internet, women are at this special moment in history poised to dominate the future.

Shlain believes language and literacy have been responsible for the subjugation of women by men throughout recent history. This is because language and, even more so, print are functions of the left hemisphere of the brain and as such essentially male-dominated functions. What is happening now, Shlain says, is that the graphical, visual nature of film, television, and, most significantly, the internet, have put women, in whom the nurturing, holistic, right hemisphere of the brain dominates, back in the driving seat, or at least back in the more equal position in society that they are thought to have occupied in the earliest human cultures. 'Misogyny and patriarchy rise and fall with the fortunes of the alphabetic written word,' writes Shlain.

Fifthly, and in what may be the most awesome discovery mankind has ever made, we may this century discover what consciousness is. There is clearly far more to mind than brain, and we will encounter arguments in chapter five that while computers will probably soon exceed human brain power, no machine will ever ask in any meaningful sense, 'Who am I?' or have a genuinely creative thought. But, with consciousness research one of the hottest topics of the moment, it seems that we are about to lift the bonnet lid on this most perplexing problem of human existence.

What we find inside *could* be disappointing, just another wiring diagram. But it could equally prove so impenetrable that we start to feel our striving for explanation of everything around us has been ultimately pointless, that we got the whole progress thing terribly wrong, and should give up science as a dead end.

That may not be the future as many futurologists envisage it. But it is incontestable that it would still be the future.

Chapter 3

GLOBAL WARNING

Environment

Futurology is a fascinating but essentially fun pursuit. It panders to our curiosity and impatience for happier times; alternatively, for those of a more depressive disposition, it provides the raw material for a gratifying dose of misery and pessimism. This element of gratification one way or another could be among the reasons why the current Pope, John Paul II, expressed his misgivings about futurology in 1998: 'No one but God knows our future and can guide our steps in the right direction,' he warned humanity. Futurology's sense of impatience may also be why the devout in Muslim countries complain to Dr Azam, of the Pakistan Futuristics Foundation and Institute, that studying the future is against Allah's will.

Big corporations, earnestly and expensively trying to worry a few clues to the future out of the available evidence, might argue that it's not a fun activity at all, but a matter of cash and employment. Yet in the bigger scheme of things, even such commercial disasters as, say, Sony backing their own Beta video format in the early 1980s when they should have swallowed their pride and taken up JVC's VHS system, or Marks and Spencer in Britain refusing to appreciate that their clothes stores had come to appear dangerously out of date, are small fry.

The single area in which we look to futurology for guidance on matters of life and death is the fate of the planet. This is hardly surprising, since a multiplicity of well-argued expert predictions, available in all good bookshops and TV documentary series, promise us a dazzling choice of imminent and excruciatingly unpleasant death. And because ecologists, environmentalists and those anxious about overpopulation are all too willing (and always have been, since Malthus and before) to offer prescriptive fixes for the latest line in disaster scenarios, we are all ears. Our appetite for disaster scenarios does not seem to diminish when rival experts disagree that a problem even exists, as invariably happens.

Of all the perceived threats to the planet's future, global warming is currently regarded as by far the most likely to come about and the most dangerous. Although, as we will see, there is a minority of informed people who think it is neither likely nor dangerous, nor for that matter our fault, global warming is almost universally regarded as the biggest issue in the whole of futurology. And as scientific theories about the future go, global warming is not only frightening, but remarkably simple and elegant.

It benefits for a start from having symptoms we are all convinced we can see and feel, and these indicators are confirmed by scientific observation. That the weather at the turn of the millennium is in a peculiar phase is clear to both layman and scientist. 'Weird' weather – violent hurricanes, floods, droughts, storms, warm winters in cold places and cold snaps in hot places – is manifestly on the increase, and the general trend of that weird weather is towards warmness rather than coolness.

Across the world, although this is less obvious to most of us, ice is also melting. 'Earth's ice cover is vanishing at an astonishing rate,' said a report from the influential Worldwatch Institute in Washington in May 2000. Every two years, Worldwatch said, the Arctic icecap is losing an area the size of Denmark. Three country-sized ice sheets in the Antarctic have also disintegrated, and reports of icebergs the size of Belgium coming free and taking to the seas have become common. The ice that remains in the Arctic, which is no more than a raft of frozen water surrounding the North Pole, is losing its thickness at an alarming rate, its overall volume, according to Worldwatch, down 40 per cent in thirty years.

Ice melting from the Arctic probably doesn't raise sea levels, as it is already effectively part of the sea and doesn't add to the volume of the oceans when it changes back to water. However, ice melting on land in this general warming up of Earth is another matter. The water produced has to go somewhere, and it seems to be running into the seas, where water levels have already gone up by a measurable ten to twenty-five centimetres in a hundred years. Glaciers across the world are contracting and are estimated to be smaller than at any time since 3000 BC. Cable car pylons in the Swiss Alps, which were anchored into soil permanently frozen with glacial ice, are coming loose and having to be shored up with concrete. Greenland's ice cover also seems to be melting into the sea by more than three feet a year.

It is possible, of course, that worrying as all this is, it may be part of a natural climatic cycle about which we can do nothing. There have been warm spells in history, such as the balmy period in the Dark Ages known as the Little Optimum, when wine was produced in northern Europe. There have also been Ice Ages and mini Ice Ages; until quite recently, the River Thames in London often froze over in the winter thickly enough for a funfair to be held on it. The majority of climatologists, however, passionately believe that this time, global warming is mankind's doing.

It is said to have started around a hundred years ago, when we began burning carbon-based fossil fuels such as coal and oil in enormous quantities. If you imagine the Earth's atmosphere as a duvet which keeps the Sun's warmth in, then 'greenhouse gases', especially carbon dioxide, sent up into the atmosphere by industry and car exhausts, have the effect of increasing the duvet's tog rating until we are uncomfortably hot. Scientists track the amount of carbon dioxide in the atmosphere by comparing modern air with that trapped in air bubbles in Greenland's ice. These measurements seem to show that there are 360 parts per million today, as against 315 parts per million in 1958 and 270 parts per million in the time of the dinosaurs.

The planet's average temperature, they report, has risen by 0.5°C (1.5°F) since 1900 as a result of all this new carbon dioxide, and the warming is speeding up exponentially, with eight out of the ten hottest years on record falling in the 1990s. Global warming, it is said, will also get still worse as the polar icecaps and glaciers melt, since the whiteness of snow and ice plays a part in reflecting sunlight back into space. As the warming effect gets more intense, with less reflective whiteness to keep us cool, we will roast all the more severely. The standard and widely accepted estimate for the rise in temperature by 2100 due to this 'greenhouse effect', assuming that we fail to cut down on our burning of fossil fuels, is said to be between 1.5°C (3°F) and 3.5°C (7°F), which would produce a sea level rise of one metre along with more turbulent weather systems. It is also generally agreed that it will take until 2050 before manmade climate change begins to become fully evident.

Even if mankind caused all this disruption, runs the manmade global warming theory, we have the power to reverse it. But it's going to be tough. Many climate researchers believe we would need to cut greenhouse gas emissions immediately by 60 or 70 per cent, which would mean a complete halt in the use of petrol- and diesel-engined cars. The best agreement we have so far come to however, at the Kyoto Summit of 1997, was for the industrialized countries to reduce their emissions by 5 per cent by 2010. But even that deal has stalled, because the United States, where less than a twentieth of the world's population produces a fifth of the world's greenhouse gases, cannot agree politically that cutting greenhouse gas emissions is necessary or desirable.

But of course global warming is far from the only cataclysmic scenario on offer in the modern age. Indeed, it is one of the milder forms of apocalyptic death being proposed for us all. Doom junkies have rarely enjoyed such an abundance of potential grief, and the blossoming of so many theories – all with accompanying little industries of researchers, institutes and disaster journalists – prompts a suspicion that while it all could be true, it could all equally be somewhat hyped up. Just three recent books and one TV special from what cynically calls itself on occasion 'the disaster industry' will serve to illustrate the point.

Apocalypse, a book by volcano expert Bill McGuire, professor of Geohazards at University College, London, promises that lying in immediate wait for humanity are: a volcanic blast sufficient to devastate a continent and freeze the rest of the world to death, a giant tsunami tidal wave powerful enough to devastate the cities along the Atlantic and Pacific coastlines, an earthquake capable of destroying the world's economy and an asteroid impact large enough to kill a billion people and reduce the remaining population to a Dark Ages civilization.

Impact Earth by Austen Atkinson, another highly respectable British science writer and broadcaster, likens our planet to the target in a cosmic shooting gallery, guaranteed eventually to be devastated by an asteroid, meteor or comet impacting with a power five times greater than the world's entire nuclear arsenal. Then there is the even splashier *The Coming Global Superstorm* by Art Bell and Whitley Strieber. Here is a global warming prediction which overturns the hope that we might be able to survive quite comfortably in a world where life has simply become more of a beach, with warmer temperatures and lots more coastline thanks to all the lower-lying land having flooded and formed thousands of islands. Bell and Strieber go instead for the (commercially sexier) *violent* global warming option, whereby the North Atlantic Current, which pumps warm water into the Arctic, heads south, spawning a sudden massive storm which relentlessly batters billions of people with icy 100 mile per hour winds before burying any survivors under tens of metres of blizzard snow.

These books and plenty more like them, many in pretty lurid jackets, have earned themselves entire shelves in bookstores. Apocalyptic television acts as a potent pre-sales boost for disaster book sales. None of these countless TV shows was more terrifying than a BBC *Horizon* documentary aired in February 2000. It suggested that any danger of climate change caused by human activity could very soon be dwarfed by a scarcely acknowledged natural hazard in the form of a hidden super-volcano, erupting from a caldera, a huge underground collapsed crater.

A caldera as described was like a massive subcutaneous boil with pressurized molten rock and debris – magma – for pus. Geologists, it was said, believed a vast caldera, most probably the one five miles beneath Yellowstone National Park in Wyoming and measuring forty-four miles by nineteen, will soon go off with a force hundreds of times greater than the Mount St Helens volcano. A caldera eruption will apparently be more like an impacting asteroid than a mere volcano in its destructive power. The blast would wipe out most of the western part of the United States, and send temperatures plunging in a devastating worldwide volcanic winter. The Yellowstone caldera, the BBC said, has erupted roughly every 600,000 years, but with the last incident 640,000 years ago – it was this which gave the world the natural beauty of the National Park as we know it today – the next was said to be imminent.

Professor McGuire was on hand to state that there have been two such events worldwide every 100,000 years for the last two million years, the Pacific Rim and southeast Asia being especially vulnerable, and that 2074 was a likely date for the next Yellowstone catastrophe: 'We're getting ready for another eruption, unless the system has blown itself out,' he said. 'But the ground surface deformation and other signs measured by satellite suggest it's still active, and on the move.' It also emerged in the *Horizon* programme that Europeans need not feel left out of the caldera scare, since there is a smaller potential supervolcano bubbling away near Naples in southern Italy.

Erupting calderas are by no means a fringe prediction, but to date they have rarely been discussed in the media or referred to by politicians. Even many ecologists haven't heard of them. Depending on your view, of course, this could amount to a cover-up on the part of 'the authorities', but the US Geological Survey, while acknowledging that the Yellowstone caldera will probably erupt again, is reassuringly – foolishly, perhaps – laid back about the real risk. 'The current rates of seismicity, ground deformation, and hydrothermal activity at Yellowstone, although high by most geologic standards, are probably typical of long time periods between eruptions and therefore not a reason for immediate concern . . . there is no indication that an eruption is imminent or what kind of eruption may come next,' the USGS says, adding that its scientists and the University of Utah 'are studying the region to assess the potential hazards to provide warning if the current level of unrest should intensify'.

Even if we were to imagine that the peril from calderas is somehow being hushed up, when it comes to global warming, there is a worldwide orthodoxy which dictates that calm-mongering be abandoned for full-on doom-mongering. The overwhelming consensus is that global warming is (a) really happening and (b) mankind's responsibility, rather than some natural phenomenon.

Yet is it at all possible that the belief in global warming has become an unthinking orthodoxy? Are we absolutely sure that even the premise, that the world is inexorably heating up, is right? It is widely assumed to be so in the scientific community, yet there are doubts among researchers.

Any statistician trying to extrapolate a million-year trend from a hundred years' numbers would be castigated for working from too limited a sample. Furthermore, the hundred or so years of temperature records on which we are forced to base the heating hypothesis are admitted, even by global warming proponents, to be patchy in reliability. Some of the latest satellite tracking of atmospheric and surface temperatures, additionally, fails to show any clear warming trend.

The very contention that a relatively puny species, a troop of glorified apes which could still fit shoulder to shoulder on to a scrap of land the size of the Isle of Wight, off southern England, has damaged, and without particularly trying to, something as vast as an entire planet will still strike many as an especially audacious example of the arrogance of the present. The twin idea that we can rescue the planet does not seem much less conceited.

There is no shortage of natural phenomena that could explain global warming. Among them are normal periodic variations in the Sun's power, natural changes in circulation of heat-bearing ocean currents, and the Milankovitch Cycle, which suggests that the Earth's orbit changes in shape periodically, and hence the planet's climate ebbs and flows with it.

Another is the possibility that we are merely rebounding from the last Ice Age and are in a natural reheating phase, following a particularly extended cold snap – a mini-Ice Age – which lasted roughly from the fourteenth to the nineteenth centuries, and itself came after the warm weather of the Little Optimum.

Yet even in the United States, with its infamous resistance to ratifying the Kyoto Convention on reducing carbon emissions, none other than Al Gore is a passionate supporter of the global warming hypothesis.

In the UK, which, being smaller and more crowded, happens to be both more gravely choked with motor traffic and better adapted to alternative and public transport, global warming is even less challengeable than in the USA. The media, with few exceptions, treats the greenhouse effect as a given, and to dispute the conventional greenhouse effect wisdom seems to be regarded as eccentric, and maybe dangerously antisocial. Scepticism on this one issue, such an admired quality in most areas of life in Britain, is more or less taboo.

And it is easy to see why. Our response to global warming is seen as a matter of survival rather than as an intellectual game. An adviser to the British Department of the Environment would brook no doubt about the subject, even privately telling me: 'It is beyond all reasonable question now that global warming exists and is caused by humankind. Please reserve any scepticism you harbour for the few remaining opponents of the theory, not for its proponents. It is terribly dangerous and irresponsible to cast doubt on global warming as some kind of politically correct flavour of the month. Anyway, you simply will not find a respectable academic voice raised against it.'

It is perhaps inevitable that global warming was not quite a new theory when it became popular in the series of sweaty summers which hit the United States in the mid to late 1980s. Baron Joseph Fourier, a French mathematician known also as an Egyptologist, was, in 1827, the first to understand that the Earth is and always was a greenhouse, kept warm by an atmosphere that reduces the loss of heat. A Swedish chemist, Svante Arrhenius, worked out in 1896 that carbon dioxide in the atmosphere increased global temperature, and that global warming could theoretically be manmade.

As early as 1957, it was theorized that human emissions of carbon dioxide could exceed the rate it was soaked up by natural means, but warnings about manmade warming went unheeded until the summer of 1988, when the fact that five out of the previous six summers in the United States had been the hottest and most drought-ridden on record began to cause official concern. But even from the start, dissidents were pointing out that most of the twentieth century's warming had occurred before 1940, when carbon dioxide emissions were much lower than they are today.

The heretical view that global warming might not be caused by man has, however, been swept aside with a unanimity only paralleled by that which once supported the theory that AIDS would kill a huge proportion of the Western, heterosexual population by 1990. That early consensus view of the development of AIDS did not materialize quite as expected, but this did not damage the reputation of those leading the consensus.

As with AIDS, global warming's champions seem to comprise the brightest and best of science. Some 16,000 Americans, a high proportion with PhD status, have signed an anti-Kyoto petition, but it is the pro-global warming advocates who overwhelmingly retain the popular perception as the good guys. There has yet to be a Nobel Prize awarded for disproving global warming, and it is unlikely there will ever be.

The pro-global warming lobby's evidence is voluminous and impressive both to fellow scientists and to the media and lay public. The evidence of those arguing that global warming is, essentially, a hoax, tends, meanwhile, to be contained in websites which have the unfortunate slickness about them of industrial sponsorship.

Persuasive as their arguments can sound, and impressive as their proponents' names and titles are, the same author names recur with worrying frequency in these writings, leading to the feeling that there are really very few formidable heretics fighting the conventional wisdom.

Perhaps the most perplexing, if unscientific, problem with global warming is simply this: we are being asked to believe that for the first time in the history of systematic thought about the future, humanity has got something entirely and uncompromisingly right – not partly right by a mixture of informed guesswork and good luck as in the examples we looked at in earlier chapters, but precisely right.

Furthermore, this futurological coming of age party just *happens* to be happening today, in our lifetime, and also just *happens* to concern anticipating not the development of some trivial gadget or convenience technology, but the survival of the very world. The embarrassing parallel with doomsday cults, and their oddly egotistical insistence every successive generation that theirs *really* is the one going to Hell in a handbasket, again comes to mind.

To date, probably the rightest anyone in human history has ever been about anything was when Bill Gates and Paul Allen predicted the boom in personal computing and how profitable it would be to make not so much the computers, but the software which powered them; if conventional science is one day proved right and the sea engulfs low-lying countries and islands, the 2,000 climate specialists who in 1996 legitimized the global warming theory by joining their predictive forces through the IPCC (the Intergovernmental Panel on Climatic Change), would have performed an act of futurological prediction countless orders of magnitude greater than Gates and Allen. Even though the problem as they see it will have been caused by human folly, recognizing it will have been by far the greatest feat of technology and imagination *homo sapiens* has achieved.

Yet even if we choose to be bold and confident enough to believe that this time, humanity has got something right, the immediate previous convictions of the environmental lobby do nevertheless cast a distracting little cumulus of uncertainty over its sincerest present protestations that we genuinely are close to the end of the world. Of course, awareness of recent green failings may in itself make us unjustifiably smug. But to rummage through the environmental panics of just the past few decades and come upon a golden oldie such as acid rain (a term first coined as early as 1852) or desertification or fluoridation, DDT or aerosol sprays, is oddly reassuring, like finding, abandoned and unwanted in the attic, that old Pink Floyd album we once thought we could never do without.

It seems cynical to state it, but fashions do move on in disaster scenarios as wildly as in diet plans. Neither acid rain, desertification nor water supply fluoridation turned out to be quite the calamity that was promised by various lobbies. In 1984, according to the *Economist* magazine, the United Nations asserted that deserts were swallowing up twenty-one million hectares of land a year. As things turned out, there was no net advance of desert at all. In 1986, the UN reported that a disastrous 23 per cent of trees in Europe were damaged by acid rain and under threat; yet by the end of the decade, the biomass stock of European forests had reportedly increased. In North America, where environmentalists had declared the forests acidified and dying, an official $700m ten-year study concluded: 'There is no evidence of a general or unusual decline of forests in the United States or Canada due to acid rain.' Only in Scandinavia, where 16,000 of 85,000 Swedish lakes are said to have become acidified, is acid rain still a major issue in the sense of being widely discussed and publicized. Fluoridation, meanwhile, once regarded as an eco-catastrophe in the making as well as the most appalling attack on individual rights, is barely heard about any more.

Before acid rain, in the 1960s, we had nuclear winter. Before that, of course, was smog and coal smoke, the pollution nightmares of their day in the 1950s. The 1920s, when the first biplanes started flying in and out of Africa, saw a huge environmental scare over the certainty, as public health experts saw it, of yellow fever becoming endemic in Europe. There were several big international commissions on yellow fever and air travel.

Scares over 'dwindling resources', meanwhile, have a busy if chequered history. A popular 1865 book by Stanley Jevons argued that Britain would run out of coal within a few years. Over a century later, it ran out of a coal *industry* thanks to cheaper coal imports and awful industrial relations, but that was a different thing. In 1914, with mass production of cars starting up, the US Bureau of Mines forecast that there was just ten years' worth of oil left. In 1951, the Department of the Interior crunched the numbers again and revised that oil supplies figure upward – to predict exhaustion by 1964.

Today, there are signs that even the theory of the hole in the ozone layer and attendant skin cancer epidemic is slipping down the charts a little, to be replaced by the up-and-coming fears about asteroid impacts and calderas, antibiotic-resistant super-bugs, toxic hospitals, mad cow disease, pesticides, flesh-eating Ebola, cellular phone brain cancers, GM foods and acid lakes (a post-acid rain worry in Sweden, where acidity has completely altered the marine biology of inland waters).

A new scare on the block known as 'Invasional Meltdown' is, according to two Canadian ecologists writing in early 2000 in the journal *Trends in Ecology and Evolution*, 'as significant as global climate change'. They say water ballast discharged by foreign ships in the Great Lakes has introduced 140 new species of marine life, from crayfish to zebra mussels, which have colonized the lakes and wiped out 90 per cent of the previously existing wildlife. Worldwide, the ballast-discharging practice is circulating some 3,000 species which were previously separated by huge distances.

What may well turn out a more obviously unpleasant form of future nightmare also concerns water. Unless the very act of prediction somehow diminishes the risk, there is said to be a very strong chance that water shortages will this century lead to wars over water. Dr Vandana Shiva, an Indian nuclear physicist turned environmental activist, predicted on British TV in 1999 that the taps in Delhi, which are already running dry, would be under lock and key within four to five years. The crisis may initially affect the poor in metropolises like Delhi and Mexico City, but according to Dr Shiva, it could soon affect the middle classes too, who can currently drink mineral water and fill their swimming pools with the stuff which might not be safe for consumption. Educated, well-to-do people, Dr Shiva said, would be 'tearing each other apart' for water and instead of wasting money on big cars, would spend millions of rupees on securing a water supply for their families.

It is actually a mistake to date environmental concern back only to Malthus, who in 1798 precipitated a flood tide of worry among the European thinking classes about matters of pollution – and, indeed, global warming.

There has for thousands of years been a degree of concern, not to say neuroticism, among intelligent people of a sensitive bent over the abuse to which it is felt we subject our planet and the strain we impose on its supposedly delicate ecosphere, the (as it is romantically regarded) gossamer-thin film of life which surrounds the globe.

When, for example, do we imagine the following lines were written on overpopulation? 'Most convincing as evidence of populousness, we have become a burden to the Earth. The fruits of nature hardly suffice to sustain us, and there is a general pressure of scarcity giving rise to complaints. Need we be astonished that plague and famine, warfare and earthquake, come to be regarded as remedies, serving to prune the superfluity of population?'

The author was, in fact, an early Christian theologian, Tertullian, writing around AD 200. It is possible to empathize with his anxiety, too. The world's population had expanded thirty-fold, from ten million or so to 300 million in a few thousand years as a result of the move from living unobtrusively off the land, as the first humans did, to the adoption of the unnatural, highly-unecological practice of agriculture. To a Roman like Tertullian, who lived in Carthage, the teeming masses homing in on the action at the centre of the world and eating its heart out must have caused the kind of claustrophobic angst visitors to Manhattan feel, wondering how one small place can possibly support so many eating, breathing and excreting people.

The British science writer Adrian Berry, who unearthed that

telling quotation from Tertullian about overpopulation, believes humans have an innate tendency to panic over imaginary dangers.

Berry, a Fellow of the Royal Astronomical Society, the Royal Geographical Society and the British Interplanetary Society, goes so far as to suggest that the degree of panic may routinely be in inverse proportion to the real peril, and that our societies may perversely feel at their least vulnerable just when optimism is least justified – and vice versa. He cites Pliny the Younger, writing in the second century of the unending glory and splendour of the Roman Empire, when it was about to collapse. By contrast, Berry points out, in 1720 it was believed that the world as we knew it had come to an end when the South Sea Bubble burst. The British Empire, in fact, had its most powerful days yet to come.

Dr Anna Bramwell of Trinity College, Oxford, tracks ecological thinking and the desire for a balanced environment back to long before even Tertullian, suggesting an underground green tradition existed in early Greek times and in the Bronze Age. 'There is no God the shepherd; so man becomes the shepherd,' Dr Bramwell writes, explaining the theory of ecology in her 1989 book, *Ecology in the Twentieth Century*.

'There is a conflict between the desire to accept nature's harmonious order, and a need to avert the catastrophe, because ecologists are apocalyptical, but know that man had caused the impending apocalypse by his actions. Ecologists are the saved.' Such a feeling of selectness, as Dr Bramwell illustrates, of being out of the ordinary thanks to your special ecological vision, often seems to fire ecological thinking; perhaps this is why it appealed to nineteenth-century romantics like Byron, as well as to twentieth-century German fascists like the animal-loving, vegetarian Hitler.

Overpopulation Tertullian- and Malthus-style (and then some) was a fear revived in the 1960s, and which has continued to come and go in various guises. Today's world population stands at six billion, and the standard population prediction for 2050 is currently running at between nine and ten billion. Ten thousand million may sound like an awful lot of people, but in

1964, a British physicist, John Fremlin, speculated in *New Scientist* magazine on population trends for the coming 1,000 years, and came up with something far more impressive.

He said that by the beginning of the twenty-third century, there would be 400 billion people, and by the middle of the twenty-ninth, 60,000 trillion, a million times today's population, who would spend their lives lying prone and eating recycled human bodies in cubicles in 2,000-storey buildings spread over both the land and the former seas. By this time, the population's combined body heat would cause extinction. At an interim point, some time in the twenty-seventh century, when population was still one sixty thousandth of its twenty-ninth-century terminal figure, space would already be so limited that all media references to dancing or athletics would be censored as pornographic.

Even if Fremlin may have been overestimating population growth a little, he was early in pre-empting the new appetite for environmentalism. The 1970s saw a flourishing of environmental thinking worldwide, with much emphasis again on the recurrently popular subject of 'dwindling resources'. This problem might have been easy to cope with, had not the obvious solution to it, nuclear power, become a spectre of equal horror. It was ironic that science's best effort at rendering fossil fuel burning obsolete terrified us because its technology was twinned with that of the atomic bomb. So resources were largely left to dwindle on.

In 1972, the Club of Rome, a group of politicians, business people and international bureaucrats, issued an influential report, *Limits to Growth*, which predicted oil running out within a decade. Supplies of natural gas, zinc and most other important minerals were also confidently stated to be a few years from exhaustion.

Dr Paul Ehrlich, a leading environmental expert to this day and a Stanford University entomologist, had already become a bestselling author. In his influential 1968 book *The Population Bomb*, he predicted a catastrophic world food shortage: 'The battle to feed humanity is over,' he wrote. 'In the 1970s the world will undergo famines. Hundreds of millions of people are going to starve to death.' In an article for the first Earth Day in 1970, Dr Ehrlich asserted that sixty-five million Americans and four billion more people around the world would die of starvation in a 'Great Die-Off' between 1980 and 1989. In 1990, he and his wife, Anne, still undaunted, published another book, *The Population Explosion*, predicting the same cataclysm again.

Well, it was an interesting idea, but according to Food and Agriculture Organization statistics quoted by the *Economist*, food production has actually risen in exact proportion with population, and per capita calorific intake has grown steeply over thirty years – even in the developing world.

Even so, the sexier, more media-friendly Ehrlich pessimism has remained impermeable to any more optimistic view, and is routinely repeated by the great and the good, in spite of the fact that such famines as there have been in areas of countries like Ethiopia since Ehrlich's prediction are all the more tragic for being easily avoidable.

The catastrophic droughts, like the frequent floods in low-lying areas like Bangladesh, are clearly real, and may be a spin-off of global climatic change. They may even be a direct result of fossil-fuel burning in the industrialized nations. Yet never is it suggested that the food does not *exist* to save the starving; it is simply in the wrong place, and is kept there by a combination of local politics and international apathy. If J.D. Bernal's belief is true that the future will consist of what we want it to consist of, we should all start wishing fervently for the Western obscenity of wasted food, with even a small country like Britain dumping 500,000 tonnes per year of edible groceries, to become a thing of the past.

Running neck and neck with dwindling resources as a 1970s *cause célèbre* there was the little matter of the Ice Age. Doom by glacier was hot, you might say, in the environmental movement's salad days. This was not the global warming-induced Ice Age predicted by Bell and Strieber's *The Coming Global Superstorm*, but a quite different one, foreseen before we realized (or perhaps just developed the arrogance to imagine) we could affect the global climate ourselves.

Concerned by apparent signs of a sharp global *cooling* between 1940 and 1975, a huge concern arose in the 1970s that the next regular, cyclical, non-manmade Ice Age, which was (and is) a couple of thousand years overdue, must be on the way, and could, as previous Ice Ages are believed to have done, take a grip within a mere few hundred years. A new Ice Age would be savage. Most of Europe, Asia and South America would be submerged by metres of glacial ice.

Even with our technological expertise, humanity would have to flee to a narrow band of land around the Equator to survive. 'A new Ice Age must now stand alongside a nuclear war as a likely source of wholesale misery and death,' said *International Wildlife* magazine in 1975. In the same year, *The New York Times* reported that there were 'many signs that the Earth may be heading for another Ice Age', while *Science* magazine in the following year believed we were 'heading for extensive Northern Hemisphere glaciation'.

Another passionate supporter of the Ice Age theory in the 1970s was Stephen Schneider, who went on to become Department Director and Head of Advanced Study Project at the National Center for Atmospheric Research. Today, as Professor in the Biological Sciences Department at Stanford University and an environmental adviser to Al Gore, Schneider is one of the world's leading global warming proponents.

The only way to show you have a mind may, as Mark Twain said, be to change it sometimes. But Professor Schneider's conversion on the road to global warming seems a little alarming. So converted was he that as early as 1992, he was telling the *Boston Globe*, 'It is journalistically irresponsible to present both sides [of the global warming theory] as though it were a question of balance.'

Raking over memories of the once trendy Ice Age theory, so very different from the global warming scenario, may well be the point at which many people despair altogether of scientific prediction.

For today's global warming dissidents tend to be as adamant in their support of a new Ice Age being imminent as is the larger and more influential global warming lobby in its assertion that we will shortly be drinking Norwegian champagne and northern Scottish clarets – and travelling a great deal by boat around the island archipelagos which were once continents.

The anti-global warming, pro-Ice Age lobby argues furthermore that, far from winding down greenhouse gas production and building sea walls, we should be planning vast space mirrors to reflect precious sunlight down on to the new, glacier-blighted wastes and warm the Earth up a little.

An interesting point made by the anti-global warming scientists in reference to their Ice Age prediction is that it is actually rather *more* apocalyptic than all but the most extreme global warming visions. Professor S. Fred Singer argues: 'There is no scientific consensus about the reality of global warming, but no one doubts that we will be seeing the onset of the next Ice Age soon. The general global cooling between 1940 and 1975 caused great concern. Then came a sharp warming between 1975 and 1980 and little change since then. I'd be more worried about a future cooling than about greenhouse warming. Adaptation to cooling would not be as simple as adaptation to warming.'

What is profoundly striking when trying to assess with some objectivity whether we even have a future worth talking about is the balance that is again and again struck between prognoses which are really quite dismal, and the belief, be it a psychological defence mechanism or not, that our adaptability and ceaseless ingenuity will, as they always have in the past, see us safely through. Even though we now have the tools and experience to predict what seems to be a grim environmental future, certain optimistic indicators worry away at the most pessimistic of us. It remains the case, for instance, even though you can buy a 'Malthus Was Right' bumper sticker in the States, that he doesn't yet seem to have been right. An increasing population has mysteriously been matched by an increasing standard of living across the world, except where war and corruption have dented the trend locally at specific times.

Furthermore, memories being short, we forget that we have fought back at our own environmental folly in the past – and won. How so? When I wondered in the last chapter what would most astonish a Victorian futurologist deposited in our world, I deliberately left out what might most amaze him not on first sight, but after a day or two. It would, in Britain at least, be the cleanness of the air from absence of coal smoke. Clean air legislation in Britain was dictated by environmental anguish and was so successful that, a little ungratefully, even those old enough to know how filthy the air was forty or fifty years ago hardly notice the difference. Anyone aged over forty or so and from the north of England can remember buildings, originally of honey-coloured stone, that were permanently jet-black because of coal smoke from chimneys and railway engines. They *can* remember this constant feature of urban life – but they rarely do.

Environmental concerns had a dramatic effect on the nuclear industry, which, post-Three Mile Island and Chernobyl, made real efforts to make its power stations safer – even though they had clearly lost the argument and the proportion of electricity generated by nuclear means is now diminishing. The same applies to cars, too. The urgent battle to remove lead from gasoline, following evidence that it was harming children's intelligence, was won over a comparatively short period of twenty years. A modern car is said to put out some 5 per cent of the pollution a new vehicle did in 1970.

That is quite an achievement, but there is more to come. A worldwide environmental upgrading of our civilization, fired by nothing more than old-fashioned capitalist self-interest, could be the catalyst for the biggest business boom of the first half of the twenty-first century. Hybrid gasoline/electric vehicles are already available from Toyota, and by 2004, we are promised hydrogen-powered fuel-cell cars whose only exhaust will be water. As they become popular, exhaust smog and global warming should wane – although there may well soon be an environmental steam crisis.

But we may even manage to escape the steam problem. Thanks to the internet and its successors, the probability is that later in the twenty-first century, we will need to fly and drive – and particularly commute – far less than we now do. The internet may be the beginning of a revolution far greater than the internal combustion engine – although the increased flow of knowledge could equally fuel a demand for more rather than less travel.

Garbage and industrial waste disposal is another outstanding area of ingenious progress. Even the messy, packaging-saturated United States is already recycling 25 per cent of its 180 million annual tons of household rubbish. When waste products start to be traded globally on the internet later this century, the concept of discarding garbage as landfill will become as unpleasantly quaint as the pre-sewerage days, when chamber pots were simply emptied into the street.

In industry, some of the money-saving and environment-protecting solutions already being put into action are quite inspiring. In Kalundborg, Denmark, *TIME* magazine reported in 1999, companies have co-operated to build an eco-industrial park, where an oil refinery, a power company, a pharmaceuticals manufacturer and a wallboard maker pool their inputs and outputs of energy, and sell their pollutants to one another. To complete this satisfactorily interlocking mini-economy, any remaining excess heat from the park is sold on to nearby houses and commercial greenhouses. Dozens of similar eco-industrial parks are being developed all over the world, according to the magazine.

There is seemingly no end to our born-again recycling fervour. In the United States, chicken feathers, which large poultry processors discard at a rate of over four tons per hour, will soon start being recycled as plastics suitable for anything from biodegradable sweet wrappers to dog bowls to car dashboards. The United States alone produces enough chicken feathers annually to manufacture an estimated two million tons of plastics.

Ingenuity and adaptability (egged on by a touch of financial self-interest) have been the mainspring of each of these advances. The fact that we have got the measure of our global warming problem and are now proposing international action to reverse it is perhaps prime among examples of that ingenuity.

For physicist Professor Murray Gell-Mann, our discovery of communal, global action could be humanity's greatest act of ingenuity and problem solving. 'Larger and larger groups of people are perceiving themselves as being "We" rather than "We" and "They,"' Gell-Mann says. 'We do have a very strong proclivity to break into small groups that don't like one another and engage in violence against one another or against the environment, but both of these are modifiable by culture.

'Something like the Montreal Convention is a very interesting example of how that can now work. The damage from the ozone hole is still relatively modest, but on the basis of predictions – and not predictions made with absolute certainty either, just considerable likelihood – governments and commercial organizations have actually responded, committed themselves to changing their behaviour. We are looking for the same sort of thing with the Biodiversity Convention and the Kyoto Convention. More and more, we are beginning to pay attention to insights about the future even though they are not perfect. It is not beyond the bounds of possibility today that governments take action on the basis of approximate predictions.'

Such cautious, provisional hope is widespread among those with the greatest breadth of vision. Niles Eldredge, a palaeontologist at the American Museum of Natural History and author of a book called *The Triumph of Evolution*, points out that interfering with the natural world is nothing new for mankind; the moment agriculture began 10,000 years ago and we took food production into our own hands, we stepped outside the natural way of things. Only now, he says, 'We are emerging from a 10,000 year vacation from nature still not fully realizing that our own survival hinges on reducing the damage we do to the Earth's natural systems.'

Even in the exciting world of taxation policy, ingenious solutions are being mooted to scale down our use of natural resources and its resultant pollution, not to mention cajoling suburbanites around the world that a four litre all terrain vehicle is not necessarily the ideal transport for a trip to the shopping mall.

A leading global warming proponent, Dr Kevin Trenberth at the National Center for Atmospheric Research in the United States, has a particularly neat and realistic policy proposal on the matter. He argues that it is vital to raise gasoline taxes, but that it should be done stealthily. 'You shouldn't do it abruptly, because that changes everyone's planning horizons and the way in which things are done. Instead, you should put on one cent per gallon tax every month, and then after ten years, it's one dollar twenty more expensive than it is now, but you know that this is going to be the case, you know gas is going to be more expensive in the future, and so you plan differently and include buying a smaller car in your planning horizons.'

All these examples of the inexorable inner will of human beings, and how it has triumphed creatively over adversity, are very fine, of course, but we are entitled to wonder how the devil even our problem-solving brains and native ingenuity could save us from a calamity on the scale of impact by a rogue asteroid, from a cataclysm of the order of the Yellowstone caldera erupting – or from global warming.

As far as calderas (or indeed regular volcanoes) are concerned, it has to be said that not even the wackier reaches of the internet can shed any light whatsoever on methods of debilitating them by such Hollywood-esque methods as spiking them with nuclear weapons. We can imagine that, perhaps, we will find a way in the future to puncture the floor of the pressurized magma chamber and release its trillions of tons of molten detritus downwards, further towards the Earth's core. But for the moment, such ideas are barely even in the realms of fantasy; and yet the solutions actively being proposed to save the world from comet impacts are as fanciful or more so than Bruce Willis's exploits in the movie *Armageddon*, in which he landed on an Earth-bound asteroid and managed to drill a nuclear charge into its core, thus saving the world.

Life imitating art, as it invariably does, scientists at Los Alamos and Lawrence Livermore national laboratories in the United States are already working on Bruce Willis-style defence scenarios, involving nuking asteroids detected years in advance by beefed up satellite-based detection systems.

Proposals for such rogue asteroid tracking systems are themselves being worked on by two Pentagon-backed research groups. Other plans suggest conventional explosives could be used to blow an invading asteroid apart, while one especially imaginative idea proposes attaching a rocket motor or solar sail to any threatening mass of rock while it is still years away from Earth and simply steering it off course. One of the many worries still perplexing scientists is that we could blow up one asteroid only to discover too late that another fragment of rock is travelling behind in its shadow, invisible to us. It may sound like a paranoid concern, but there is certainly a movie treatment in it.

Ambitious fixes are also being dreamed up for global warming. The most dramatic, known as geo-engineering, can be equated to fitting the planet with a permanent pair of Raybans. Small particles, it is said, could be placed in the upper atmosphere to reduce the amount of sunlight reaching the Earth's surface. Less dramatically, we could simply batten down the world's hatches against the rising oceans, building better sea defences, moving populations and crops on to high ground and perhaps living more on massive ships and manmade islands.

Another, less spectacular form of human ingenuity in the face of stacked odds against – but one in which there most definitely *isn't* a Hollywood storyline – may concern our having to adapt our future meal plans. Traditionally, all over the world, human beings have marked their ascent from poverty to middle-class status by moving from a grain and pulse diet to hogging out on beef and pork. If we can convince ourselves to move down the food chain a rung, however, and wean the species away from eating animals as a matter of daily routine, it seems that we could benefit the planet in a variety of ways.

Firstly, meat production is massively wasteful of water; a

pound of beef, it has been calculated, requires nearly 700 gallons of water to produce. An equivalent amount of similarly nutritious plant matter takes a seventh of the water to grow.

Secondly, meat production is hugely polluting, with livestock excreting 130 times as much waste as people; a giant pig farm in the United States can apparently produce more sewage than the whole of Los Angeles. Thirdly, meat in large amounts isn't even good for us; China and Japan, where meat eating has increased exponentially in a few decades, have been visited for the first time by mass obesity, heart disease and cancers of the breast and lower alimentary tract. That is allowing for what might be called 'organic' meat; the effects, if any, of eating flesh imbued with antibiotics and hormones, or even genetically modified, have yet to be agreed on.

A mass move towards vegetarianism in the next century may also be accompanied by a revival of a slightly out-of-date futuristic dream – the farming of the seas. Nearly 20 per cent of the world's current human population relies on fish for its protein, and the recognition of its qualities as a healthy food promises to push that proportion ever upwards. It is a truly bizarre quirk of humanity, then, that in the 5 per cent of the planet's biosphere which is land, we artificially produce animals (at massive cost and effort) for food, while from the remaining 95 per cent of the world, the oceans, we extract seafood on the wild game-hunting principle of killing whatever nature provides and then moving on to the next hunting ground. This has not unsurprisingly led to the near extinction of the breeds we like to eat, as well as to ridiculous prices being commanded for fish. A portion of cod is now more expensive than porterhouse steak in the UK, while a whole tuna can cost as much as a new small car. Yet if instead of growing them, we had to hunt beef cows in the wild, they might soon be rare enough to command the price of a Learjet.

Fish farming or aquaculture, then, which is slowly taking off around the world, could, as was predicted decades ago yet curiously forgotten about, be a massive new source of food in the coming centuries. Arthur C. Clarke, in his books *The Deep*

Range and *The Challenge of the Sea,* came out as a keen advocate of seafood farming. The only difficulty with these predictions, made in 1957 and 1960 respectively, is that they envisaged the beef cattle of the sea being whales, a reflection of the era Clarke was writing in, before we came to believe that whales are as clever as us and a good deal better behaved.

Nonetheless, Clarke's vision is worth recalling, since it could presumably be adapted to another less emotionally correct form of marine life, either plant or animal. He described sea ranches thousands of miles across, run by international bodies and with nominal borders, so they could move with the seasons from the polar regions to the Equator and back, wherever the whales could find better plankton grazing. 'The life of a twenty-first-century whaleboy,' he wrote, 'would still have plenty of excitement, even though he may use an atomic-powered submarine instead of a horse and curtains of electric pulses instead of a lariat.'

All the examples of human resourcefulness we have looked at here, of course, are fundamentally selfish in that they concern the comfortable survival of our own species.

But along with that quite miraculously increased sensitivity of individual human cultures to the right to life of other cultures, another extraordinarily futuristic development has taken place – a belief in the right of other species to exist, and an acknowledgement that as the leading fauna currently around, it is our duty to do whatever we can ingenuity-wise to ensure what is fashionably called biodiversity – even when it is manifestly our fault that thousands of species have died out, and are continuing to do so.

The first examples of the human will to extend an altruistic paw to endangered species are no more than a hundred years old. Before that, exotic animals were treated largely as a sadistic entertainment. At a menagerie in the Tower of London founded by King John in the thirteenth century and only closed in 1835, lions and tigers were caged together to encourage them to fight, while elephants were fed broken bottles and ostriches' nails – apparently for no better reason than to cause them pain.

But by 1822, the Society for the Prevention of Cruelty to Animals was being set up in England, and soon after it was founded in the mid nineteenth century, the Bronx Zoo in New York was successfully reintroducing bison to the prairies, where they had been nearly made extinct by hunting. Fifty years later, the zoo buried a time capsule to celebrate its half century. When it was opened in 1999, the words of a curator were read, expressing the hope that one day it might be possible to breed gorillas in captivity. In fact, by 1999, forty-three gorillas had been born in that zoo alone.

Today, alongside the often gruelling struggle to preserve species in the wild, ever more imaginative suggestions are being put forward for promoting biodiversity. Artificial insemination has been performed successfully even with pandas, a particular achievement since they only ovulate once a year. Cloning is currently a matter of high controversy. On the one hand, it could undoubtedly ensure an almost inexhaustible supply of what are wryly called the 'charismatic megafauna' – the sexy animals like pandas, tigers and gorillas. On the other, it could be regarded as pointless, since it keeps the gene pool completely stagnant, when genetic diversity is valued above all by zoologists. On the other hand still, cloning could bring back extinct species. 'Can you imagine what having a small herd of mammoths would do for zoo revenues?' speculated palaeontologist Larry Agenbroad of North Arizona University to *Newsweek* magazine late in 1999.

It inevitably falls to science fiction writers to stick their necks out the furthest on the future, and one of the most in vogue and respected writing today – regarded by some as a new Arthur C. Clarke – is an American, James Halperin. Halperin, a high school maths geek turned Harvard psychology and philosophy dropout, believes that in the future, human beings will live for ever – and 'nature' may just have to fit in with our plans.

'People don't want to live for ever now, but that's only because we weren't brought up to expect that, and it's hard to deal with. But for people who were born in an age when it's expected, death will become an intolerable concept. I think people will recognize their feelings about that when they no longer have various forces in our society telling them that death is part of the natural way. Of course it *is*; but nature is abominably cruel, and we are very close to having the power to override it. What's so good about nature? I'm glad that it existed at one time, but I'm kind of sick of it now. It's outlived its usefulness.'

Halperin is, perhaps, being deliberately provocative when he questions the primacy of all that is natural, but there could be a point here. Is it not a case of tough luck for disappearing species if they happen to have lost out to us in the evolutionary battle? Also, if it's right for us to be proactive in helping the survival of the not-so-fit, would that include disgusting creatures such as cockroaches or even viruses if they were threatened with extinction, or does biodiversity only extend to box-office worthy megafauna?

I put these questions to Dr Alan Dixson, San Diego Zoo's director of conservation science. It's possible, agreed the Nottingham- and Birmingham-educated zoologist, who has also worked at London Zoo and as director of a Primate Centre in West Africa, that we could survive as a single species and let our own population continue to expand.

'But if we impoverish the world around us and put our children and grandchildren in circumstances that we don't have to put up with, is this really the right way to go?' he asked. 'We have to link the growth of the human population and its use of resources with what is happening with animal life, because assuredly, the destruction of habitats and rainforests and the decrease of so many species are linked to our own increased use of the planet. It is not happening in some separate universe.'

It might equally, he said, be tempting for politicians to say that war is inevitable, there have always been wars, so let's have another. 'But I do think it is a huge abrogation of human responsibility to take that kind of view. In the past, there have been great formal extinctions, but those have been due to naturally selective forces. This is not natural selection; it is abnormal selection via the agency of man, which otherwise would certainly not be occurring. It is very, very unusual and has never happened before in the history of the planet.'

As for cockroach preservation, Dr Dixson, no great fan of the roach, seemed a little relieved that it was unlikely ever to be a problem for him. 'My feeling is that when we are all gone, all there'll be will be ants and cockroaches – they are incredibly hard to get rid of,' he said. 'But philosophically, lesser creatures are worthy, and behind the scenes, most of us are concerned about many other things than the pandas and so on which capture people's imagination.

'No matter what we do, I think we are going to lose an enormous amount of biodiversity, but if we do nothing, there will truly be a catastrophe, a complete loss of rainforests and the megafauna will be gone. But due to the activities of a huge number of organizations and individuals, there are extremely hopeful prospects. A lot of ecosystems are being protected, so although they are going to be whittled away, there are still going to be some rainforests in the Amazon and pandas in China. So I don't think there will be a complete disaster and the only reason there won't be is because of the effort people are making. There's no point in being depressed. We have to fight on and do the best we can.'

Of course, Dr Dixson speaks of the *disappearance* of diverse species, but evolution could equally have a way of producing new species in the future, with or without our disruptive presence. A British TV series, *The Future is Wild*, being made in 2000 with finance from The Discovery Channel, speculates with the help of scientists, graphic designers and animators on the fauna of a post-*homo sapiens* world two million to a hundred million years from now. Cockroaches, as Dr Dixson suspected, figure large in this world of giant beasts; they are envisaged as growing to eighteen inches long.

The rest of the cast of this scary proposed world seems to borrow as much from Pokémon monsters for its design inspiration as to evolutionary biology. There will also, it is said, be mutant ants which can morph to any shape, vampire bees which suck the blood of small mammals and five-foot-long flying squid which use jet water propulsion to leap through the skies and capture and eat flying dogs. The American plains will be dominated by sixty-five-foot-long creatures called (or more correctly not called, since there won't be any humans to name them) rattlebacks. These will have evolved from racoons and have a shell to retreat into to survive the severe storms forecast for the thousandth century or so. Then there will be the ratch, a giant rodent adapted to seek out dead animals from mud, and the flamingo cat, which will have developed long wading legs and have fur coloured pink by its shrimp diet.

'Our ancestors survived the last great Ice Age,' comments Dr Roy Livermore, a palaeogeologist at the British Antarctic Survey and one of the programme's advisors, 'but whether civilization will survive the next is another matter. Mass extinction could have happened before that, and new creatures would have evolved by then.'

Like eternity and infinity, trying to cope with imagining a world in which we are extinct and racoons the size of double-decker buses stalk the Great Plains is almost too much for our brains. For the moment, we have to pretend we are still in the game. But ultimately, how we – the public and policy-makers – choose to treat our planet while it still is partly, at least, ours, hinges on a combination of personal psychology, cultural factors, how we interpret the advice of scientific experts – and in the final resort, politics.

The evidence suggests that these are muddied waters, in which various eddies and currents flow to stir up the murkiness further. In the psychological whirlpool, there is one stream of belief that seems to get a certain charge from seeing the dark side of everything, which likes to appear whey-faced and miserable and pessimistic in all circumstances because optimism looks un-intellectual while pessimism is kind of cool. Another counter-current refuses to see the potential disaster in anything and is absurdly optimistic, seeing every half-empty glass as half full.

The odd thing, psychologically, about an apparent mortal

peril like the Yellowstone caldera is that the very people predicting that it will destroy the entire Western seaboard of the
USA and bring darkness down upon the rest of the world are
living with their families right under it.

If they are so confident of their data, is it fatuous to ask why
don't they kill themselves and their loved ones painlessly now
while it can be done relatively pleasantly? I put this question to
volcanologist Professor Bill McGuire, author of the frightening
book *Apocalypse*: 'I am afraid that deep down, like everyone
else, I don't really feel that any of these global catastrophes will
happen when I am around,' he explained. 'A common human
response that keeps us all, or most of us anyway, sane, and it's
statistically true too. Like everyone else I spend more time
worrying about my next deadline, or the mortgage and am
likely to be almost as surprised as the next person if an asteroid
hammers into the Earth a week next Friday.'

Culturally too, the same scientific data can resonate in different ways depending on where it is received. An intriguing
1999 survey by Dennis Bray and Hans von Storch of the Institute
of Hydrophysics in Geesthacht, Germany, of 1,000 climate
scientists in Germany, the United States and Canada showed
clearly how the north Americans, even though they agreed
global warming was a problem, were less convinced by the need
for social and political action to do anything about it than were
the Germans.

The Canadians were less inclined than even the Americans to
believe climate change would have a detrimental effect on their
society, and were more likely to feel global warming might
actually have positive effects. 'This incompatibility between
the state of knowledge and the calls for action suggest that,
to some degree at least, scientific advice is a product of both
scientific knowledge and personal persuasions, suggesting a
socio-scientific construction of the climate change issue,' the
researchers commented. A guarded way of suggesting, perhaps,
that of course Canadians don't mind the world warming up a
little, because it's so damned cold in Canada.

Peer group pressure on scientists not to rock the consensus

boat is also a factor in a subject like global warming. One prominent British physicist, who declined to be identified, told me in an Oxford pub that both he and most of his scientific colleagues across several disciplines regarded the manmade global warming hypothesis as 'arrogant hogwash' which grossly overestimated the abilities of mankind – but would never say so publicly because to do so would be professional suicide.

His refusal to 'out' global warming was more than mere cowardice, he argued. The unnecessary (he believed) measures being demanded for countering the non-existent (or, if it exists, perfectly natural) warming happen to be extremely desirable and beneficial to the pleasant and civilized progress of society, so it would be thoroughly irresponsible to argue against them.

'Those are very common points of view in science,' commented Henry Bauer, professor emeritus of chemistry and science studies and dean of arts and sciences at Virginia Polytechnic Institute and State University, as well as editor-in-chief of the *Journal of Scientific Exploration*. 'Scientists, like other people, are sheep-like. I've just been reading a lot about AIDS and what causes it. There's no good evidence that HIV is the cause, yet hundreds of thousands of medical scientists won't listen to any other possibility.'

For Professor Bauer, author of a book called *Scientific Literacy and the Myth of the Scientific Method*, the fatal combination is the scientist with a new theory and unclear of his facts – and the expert-hungry news media.

The point made by my anonymous physicist that, in effect, hypocrisy should be allowed full rein, and we should put anti-global warming strategies into action even if we don't really believe they are necessary, is more commonplace, and more cogent, than one might expect. If scientists are unable to reach a consensus on global warming, neither can the rest of us, the argument runs, and that implies that we should energetically pursue the reduction of greenhouse gases as an insurance premium against an entirely unacceptable risk.

Perhaps the most forthright argument to date from academia for a little PR-spinning in a good cause was made by the former

Ice Age enthusiast-turned-global-warming-advocate, Professor Stephen Schneider. He admitted to *Discover* magazine in 1989 that the public had to be subjected to scary scenarios and dramatic, simplified statements to drum up backing for ecological action. 'Each of us has to decide the right balance between being effective, and being honest,' Schneider said.

In the end, perhaps both our assessment of the environmental risk we pose to ourselves as a species and the determination of what we should do about it for the future belongs in the realm of politics.

There are at least two cross currents in political waters. In one direction, flows the belief that anything created by the West and by industry is inherently evil; against it rushes a current which carries the message that any suggestion that the West should alter its ways, that humanity ought to treat the planet more gingerly, is a form of covert socialism, of big government pushing the good, independent, common-sense guy around, and should be rejected.

The environment issue is, at a deep level, a left/right thing. If you are broadly of the left, believe man is perfectible if he can begin to conform to stringent, communitarian rules, but are pessimistic about the way things will work out, you will be more likely to believe the planet is indeed doomed to be killed off by our own foolishness. Part of that mindframe is also to believe that we can think and analyze our way out of trouble.

If you are politically of the conservative right, believe that man is essentially flawed but are optimistic that, so long as we leave people alone to do what they like, everything will probably work out OK, you will be sceptical of global warming. If you are of that mind, you may well also believe that it's simply too much of a coincidence and an arrogance that *we* should happen to be the generation which witnesses the end of civilization; you may even feel instinctively – irrationally, some would say – that destiny won't allow us to kill ourselves off.

Use of the term 'left' in this model doesn't always conform to political norms, it has to be said, because fascists are strangely drawn to ecology and can become terribly heated on libertarian grounds about issues such as water fluoridation. But broadly, the left/right distinction holds. It is for us to judge whether the intellectual, slightly arrogant, slightly neurotic, rather bossy 'left' will be proved correct; or whether the over-cynical, lacking-in-imagination, complacent, over-superstitious 'right' has the best tunes. That ultimately depends on whether, on balance, we feel the left has the best track record in being correct on big issues, or the right.

How can we reasonably take our bearings and estimate what our planet's destiny will really be? Is it too lazy and trite to suggest that the truth will most probably lie somewhere in the middle, where it usually does, that environmentally, things will work out far better than the worst predictions and a little worse than the best, with plenty of space left for surprises – an earthquake here, a devastating volcano there? Are we fooling ourselves when we feel instinctively that the likelihood is in ten, fifty, five hundred and one thousand years from now, we'll still be here, that everyone who predicted disaster all those years ago was kind of right, but also kind of wrong?

One example of the kind of ingenuity which tends to get mankind out of sticky patches, but that I did not mention earlier because it is less a case of wilful inventiveness than just the way things work out, is the slow but relentless growth internationally of the middle class.

Social climbing may yet prove humanity's salvation; with superior financial status, people do like to consume more environmentally costly foods and drive cars; but their cars are more likely to be new, clean models than polluting old ones. And at the same time, people also, crucially, tend to have smaller families as their economic wellbeing improves. In Bangladesh, interestingly, there has been an enormous decline in family size over a couple of decades, from an average of 6.5 to under four. That still represents a sizeable population growth, but is an encouraging indication of the way things could go.

The plentiful supply of information, and the enlightenment which comes with that infinitely available resource, is a key factor in the *embourgeoisement* of populations. It is not hard, then, to imagine the world really could be on the path to healing itself when you read on the business pages about Ghana's internet-using population rising from 5,000 to 50,000 in three years, or that Gamtel, the communications company in neighbouring Gambia, plans to extend phone lines through a fibre-optic cable network to every village in that country. 'Soon, even in the classrooms in villages, children will have access to the internet,' says the managing director of Gamtel, Mr Bakary Njie.

As always seems to happen in futurology, real progress comes about stealthily, crabwise, in the one, surprising way nobody was quite expecting.

Chapter 4

DON'T LOSE YOUR HEAD

Human Body

One of the earliest, most Frankensteinian, visions of the future of medicine involved the transplanting of human heads. This grisly procedure was the subject of some isolated experiments for a while, but then became a science fiction staple.

Today, however, head transplantation, the one spare part surgical procedure in which it pays to be the donor rather than the recipient, is being talked about again. Robert J. White, professor of neurosurgery at Case Western Reserve University in Cleveland, Ohio, wrote in a 1999 *Scientific American* article that he and his colleagues have already taken the first steps towards human head transplantation by developing pumps which would lower the temperature of blood in the to-be-transplanted head to 10°C (50°F). This cooling would help the head shut down for the necessary hour or so while it was being reconnected.

Professor White, who envisages the operation as a treatment for people who have been paralysed, successfully transplanted the heads of rhesus monkeys from one body to another in 1970. Such monkeys lived as long as eight days fully conscious and with most of their bodily functions working. The best previous attempt at such engineering had been by a Soviet scientist, Vladimir P. Demikhov, in the 1950s. Demikhov produced a notorious 'two-headed dog' that survived for twenty-nine days by grafting the upper body and forelimbs of a mongrel puppy to the neck of a bigger dog. Before that, in 1908, Charles C.

Guthrie, a young vascular surgeon at the University of Chicago, successfully attached the head of a small dog onto the neck of a larger one.

Head transplants are at the more obviously spectacular end of the spectrum of medical advances that look likely to appear over the next hundred or so years and beyond. But, as the procedure's dated antecedents suggest, the principle of a head transplant would not have seemed all that alien to the knife-happy barber-surgeons of two hundred years ago; at some stage, even with his hi-tech equipment and as yet undiscovered method of connecting two parts of a spinal cord together, White will be obliged to saw the heads off two human bodies. Not an overly elegant procedure, some would say, but just the kind of advance that forward-thinking people of a hundred years ago might have expected by the twenty-first century.

Other examples of what we thought the future of medicine might bring include the bizarre 1950s belief that 'food pills' would soon sustain us. Even if it is arguable that McDonald's hamburgers *are* those food pills, that idea has never shown a glimmer of taking off. Then there was that most futuristic of notions, cryonics, the deep-freezing of bodies in liquid nitrogen pending some unknown medical advance still further in the future which will supposedly both thaw you out and cure whatever it was you died of.

Cryonics was first popularized in 1962 by Robert Ettinger in a privately published book called *The Prospect of Immortality*. He founded the non-profit Cryonics Institute in Clinton Township, Michigan, where to this day, his mother, Mae, is among thirty-one bodies stored in a state of cryo-suspension. She died (although cryonics adherents don't approve of that term) in 1976, but one of her freezer-mates, a Dr James Bedford, was suspended as early as 1967. Both, of course, might find themselves one day revived in a world they don't understand and where nobody knows them. The possibility of a number of God's frozen people being warmed up at some future date is certainly an intriguing one for science fiction. One of the issues which might be fruitfully explored is whether these people's souls and memories are somehow deep frozen with them.

But the frozen community does not look as if it will ever be a very large one, for today, even if cryonics is not quite out in the cold, neither has it set the world on fire. In the 1960s, it seemed self-evident that cryonics simply *was* the future. Ettinger's book became a worldwide bestseller and in 1969 there were estimates that billions of us would be frozen by 2000. There are today four other cryonics firms, the leading one, Alcor, in California, having thirty-five 'patients' and around four hundred volunteers for the freezer. Sufficient computer and software denizens had signed up as potential freez-ees by 1998 for a local newspaper to declare that cryonics was 'a new Silicon Valley trend'.

Nobody, of course, has any idea if frozen people will ever be able to be thawed and revived. Several individual pieces of tissue and some human embryos have reportedly survived freezing, but the prospects for the brain remain a key problem. Some cryonics companies (but not Ettinger's) nevertheless offer 'neurosuspension' – the freezing of a patient's head alone. Ettinger, now eighty-two, disapproves of this, pointing out that it is too cruel to ask relatives to allow their dear departed to be beheaded. He has no fears of the freezer himself, however. 'As for the risks of putting yourself in a deep freeze,' he said in 1999, 'life has always been risky, and now even death is risky. But any reasonable reading of history suggests that, by and large, the future will be better.'

Head grafts, food pills and cryogenics aside, the history of futuristic thinking about medicine has been markedly less varied and exotic than about almost any other subject. There is a good reason for this: medicine has only ever had one goal, the improvement and extension of human life. How that might be achieved has changed through the centuries, but there has never been much demand for radically different goals, such as hopes for a shorter or less comfortable lifespan. In Victorian Britain, the radical concept of sanitation and public health was seen by a few conservatives as a development which would sap lower-class people's morale by making life too easy for them; the same misgivings were voiced by some when Britain started a free health service in 1948. But as a general rule, we have always striven for a better, longer life.

Of course, until the nineteenth century, few people consulted doctors, so medicine was not even the primary route for seeking a longer life. 'Physicians were a very small group of people who mainly practised on the elite,' observes science historian Dr Tim Boon of the Science Museum in London. 'The vast majority of the population would turn to the church or alternative religious movements rather than to medicine for their longevity and hopes for the future.'

But when it comes to techniques for prolonging life, people have been all too willing to embrace the latest techniques in the hope of forestalling death. At any time, doctors and patients have had faith in whatever is on offer from medicine, be it trepanning, leeches, bleeding, emesis, purgatives – or organ transplants. The faith we are prepared to place in the medicine of the moment can sometimes seem a little tragic in retrospect; Dr Boon has seen archive footage of a 1935 American newsreel, *Men Against Cancer*, in which it is confidently proclaimed that thanks to modern techniques, the disease has been all but beaten.

Such unfulfilled medical bombast, enhanced by the odd way we treat doctors as people to ask about even emotional matters in which they have no training, has resulted in an anti-medical backlash. Today, thanks to this cumulative lack of faith in modern medicine, there is a medium-to-large-scale semi-rejection of the latest and also the most futuristic that medicine can offer in favour of ancient systems from countries like China; but none of these 'alternatives' is backward-looking for the sake of being so. Homeopaths, herbalists, even faith healers may promote ancient techniques, but they do so in the cause, as they see it, of progress. They feel they are more futuristic than standard medicine, and when they read about head transplants in *Scientific American*, it is not hard to see their point.

Even though medicine's goals have been consistent, attempts to forecast the future of medical techniques have been reliably off-key, as we saw with cryonics. Such dead ends suggest, as ever, that any present confidence we have as to where medicine is headed next could be equally mistaken. 'Most of the underpinning beliefs we hold about medicine are less than 200 years old. So if we reflect on 200 years in the future, our entire understanding could once again be radically different,' says Tim Boon. 'We can't really tell which of the current fundamentals of our understanding – Freud, Darwin, DNA – will last.'

There is already ample evidence of Boon's theory that medicine could swing any way other than the expected, for the most likely coming medical advances will be both scientifically more radical and physically less messy than head grafts or defrosting grandpa. They will centre on two mega-concepts, each – in keeping with the future's habit of surprising us – totally unforeseen as little as thirty years ago. Both these big new medical ideas are in their infancy and come under the broad heading of biotechnology; the first is genetic engineering, the second, the merger of biology with electronics.

Along with offering changes to humanity which amount to little less than the re-writing of the *homo sapiens* users' manual, these new technologies will also open up a range of new problems for us to deal with. These will include such minefields as the psychological and societal implications of extremely long life or immortality becoming routine; the increasing divergence between amazing medical treatments available to the wealthy and rudimentary healthcare the poor all over the world are liable to be left with; and the formidable complications to our personal relationships of sex becoming unnecessary for procreation – and virtual sex over the internet becoming a reality.

The bio-electronic – or bionic – area is the sleeping partner of these emergent medical technologies, since at the moment it is less developed than headline-grabbing genetics. Nevertheless, the kind of electronic aids to human life currently being seriously discussed and developed are quite dazzling both in their level of ambition and daring, as well as their sheer number. Some of the following paragraphs may, consequently, read more like the ideas from a Hollywood script conference than the stuff of real science; it is all strictly non-fiction, however.

In a fifth-floor lab at the Massachusetts Institute of Technology, for instance, engineer John Wyatt, after twelve years' work with Harvard Medical School neuro-ophthalmologist Joseph Rizzo, is testing in patients a surgical electronic implant for damaged retinas. Wyatt's aim is no less than to develop an electronic retina for the blind.

At Glasgow University, scientists are working on a fifteen-millimetre-wide capsule which contains an entire miniature laboratory designed to detect diseases while travelling around the body, especially the alimentary canal. The 'lab-in-a-pill' includes a mini video camera to relay pictures from inside the body. 'We thought of measurements being taken in the most remote places, such as in space,' Professor Jonathan Cooper of Glasgow's Bioelectronics Research Centre says of the development of the tiny laboratory, 'but then we thought the most challenging environments could be the ones closest, inside the human body.' Elsewhere, teams like Cooper's are working on electronic implants to bring hearing to the deaf, and smell and taste to those who have lost these senses.

At the Royal Sussex Hospital in Brighton, cardiologist Professor Richard Vincent predicts the appearance soon of a generation of smart heart pacemakers, able to revive patients who suffer heart attacks and automatically phone for an ambulance, giving the driver an exact location by satellite guidance. These deluxe pacemakers, Professor Vincent says, will also respond to their owners' voice; you will be able to warn your heart if you are about to go for a jog or some other strenuous activity.

Still further ahead, but perfectly likely, we are told, within thirty years, will be electronic connections between computers and the brain. 'By linking directly to our nervous systems, computers could pick up what we feel and, hopefully, simulate feeling too so that we can start to develop full sensory environments, rather like the holidays in *Total Recall* or the *Star Trek* holodeck,' says Ian Pearson, a leading British futurologist who monitors technological developments around the world for the phone company, BT. Speechless brain-to-brain contact via such electronic/brain interfaces as Pearson describes is expected by around 2030.

The awesome implications of hardwiring electronics into the brain for the way we think and take pleasure will be examined in the next chapter. But such inventions, for Pearson, will be 'the beginning of a long process of convergence that will culminate in a fully electronic human being well before the end of the twenty-first century'. For Professor Richard Gregory of the Department of Experimental Psychology at Bristol University, the same prospect can be expressed in a subtly different way; he wonders whether by the end of the present century it will be clear 'whether we will be people or robots'.

Caring for parts of our bodies other than the brain will also be done in the near future by a flotilla of electronic gadgets. Domestic lavatories which perform a urinalysis and stool test, searching for any worrying trace of blood or defective DNA, are expected to become common. Toothbrushes and hairbrushes will silently test our DNA and warn us about any oddities they detect. We will carry our medical records on smartcards, enabling any doctor in the world to access our entire medical history.

Then there are nano-robots. Molecular nanotechnology (*nano* being Greek for dwarf) is the most unlikely-sounding of all the new medical technologies, but one which is nevertheless on the drawing board of a host of biotech companies.

Nanotechnology is the manipulation of material at an atomic level by microscopic machines. First proposed by the celebrated American physicist Richard P. Feynman, it holds out the possibility that one day within a hundred years or so, our veins, capillaries and very cells might be patrolled by trillions of microscopic, sub germ-sized machines programmed to clean arteries, detect cancers, even perform cell-level surgery. Other nanobots may carry oxygen to supplement red blood cells in the event of a heart attack, thus keeping tissues alive until help is at hand. All the nanobots will constantly report their inner body findings to a miniature mainframe computer lodged discretely under our skin and accessible by doctors, either in person or via the internet.

It all sounds preposterous, yet engineers at major universities have already built microscopic working machines and gears, even molecular motors a hundred or fewer atoms in length. When the Raquel Welch movie *Fantastic Voyage* came out in 1966, with its story of a team of doctors scaling themselves down to explore a patient's body in a miniaturized boat, it was derided as too silly even for science fiction; yet within fifty years of its release, *The Fantastic Voyage* may be seen as a dated but rather perceptive piece of futurology, perhaps even a bit understated.

It is the impact of scientists' growing understanding of our genes and how genetic instructions shape a human being that has come, nevertheless, to dominate medicine in the immediate millennium period.

In terms of media coverage, genetics is the global warming of medicine. Almost every day, we see media stories about the discovery of a gene responsible for anything from specific diseases, to longevity, to joke-telling. Even the business pages are full of reports of commercial companies attempting to file patents for gene sequences they claim to have decoded. And then there is the growing global angst over genetically modified foods.

Plant and animal breeders have, of course, used a knowledge of genetics for more than a hundred years. So in that we have long known how to breed a better vegetable or spaniel, genetic engineering is slightly less futuristic and scary and Frankenstein-like than some of its modern critics make out.

Of course, genetic *modification*, the idea of playing around with human genes from the inside (as opposed simply to being choosy about whom you or your cauliflowers have sex with) moves genetics forward a vital stage. It is probable that by 2025, scientists will be regularly cloning human beings to order in the lab, and offering parents the chance to choose the sex, hair colour, intelligence, muscularity, susceptibility to mental illness or whatever of their unborn children before having them grown in an artificial womb. Our gut feeling against such a prospect is doubtless coloured by the fact that dreams of 'playing God' in this way have excited political madmen for much of the twentieth century.

Meddling in any part of the process of creation of life is indeed an extremely sensitive area politically. In the 1930s, when it was first mooted that a procedure common today, in vitro fertilization, might one day be possible, geneticists thought everyone would suddenly want to get George Bernard Shaw or Lenin to 'father' their test tube baby.

As things have turned out, attempts to sell 'genius' sperm have never really taken off in a big way. But the Nazis were entranced by the idea of 'eugenics', of improving the human stock so that everyone might, presumably, *not* look like Hitler. So dogmatic were the Soviet communists, at the same time, that genetics couldn't and shouldn't be practised even in agriculture, that in 1940, they arrested one of their top plant geneticists, Nikolai Vavilov, for merely being a geneticist, and sent him to a prison camp for 'belonging to an English-backed rightist conspiracy and sabotage of agriculture'.

Today, genetics is again a hot potato. Above all, there is a persistent fear that some maverick scientist somewhere will sooner or later build a cloned human being with a malevolent purpose. The creation of such a being would leave the rest of us, were we able to contain him, with the deeply disturbing problem of what to do with such an individual – for an individual with individual rights he or she would presumably be. How would a cloned Saddam Hussein, for example, need to be treated? Would we be justified in imprisoning him for being his genetic father's clone?

At times, it seems that the real code to be broken in the furore surrounding genetic engineering is the one which will help decipher reality from hype. Is genetics really the master concept that will revolutionize medicine? Can geneticists actually deliver what they promise in terms of fine-tuning human capabilities and curing and preventing disease, or are they hyping their case to keep their funding up? And if they can deliver what they promise, do we want them to?

With genetic engineering, as with global warming and, in the next chapter, artificial intelligence, we are being asked to believe that we, one generation out of hundreds, are lucky enough to be present at the very moment of *the* breakthrough, *the* day when we finally understand how living organisms work, when all becomes clear and under our control. It's perfectly possible and there are good reasons, as we have seen before, why it might be the case; but, as one Japanese futurist, Rei Uda Kawashima, puts it and others echo: 'What we generally find is the more we know, the more unknown things we have.'

In other words, there is a distinct chance that by unravelling the mysteries of the human genome – the name given to the entire complement of forty-six chromosomes contained in every human cell – we will raise more questions than we answer. To use a parallel with computer software (which when explaining genetic engineering is unusually apt as parallels go), if we were trying to learn Windows 2000, would it help or hinder us to see the entire machine code for the monster-sized program on a miles-long printout? The answer for almost any user of the program would have to be 'no'; such raw data on its own isn't always very helpful just because you have it. The commercial companies stampeding to patent genes will often be staking their claim, gold rush-style, on worthless DNA territory; for, confusingly, not every piece of DNA that looks like a gene is a gene.

But let us first try to define terms; although genes, genetic engineering, chromosome and genome are all glibly referred to every day in the media, they are fiendishly tricky to explain. Genetic engineering (also known as gene therapy) is the manipulation, either with drugs or by physical intervention using electricity or carrier viruses, of the structure of DNA, the chemical deep inside a cell which holds within it all the 'software' needed to run an individual's physical and personality traits.

Human DNA exists in a tightly-coiled three-foot-long molecule, which forms the heart of a chromosome. These chromosomes reside in their sets of forty-six (twenty-three bundles of two each) in the nucleus of almost every one of the body's 100 million million cells. And each individual chromosome DNA strand includes within it several hundred to several thousand sub-sections called genes.

Each gene, of which humans have about 130,000, seems to dictate (or at least, have a significant role in) a particular way in which a cell works. It does this by expressing chemical orders for a specific function to be performed. Although our individual chromosome set – the genome – is the same throughout the body (apart from in the sperm and egg cells, which have half a set), genes only carry out their specific job in the area of the body they apply to. So it matters no more that a gene involved in some liver function happens to be resident in the big toe than it concerns a poet that his Microsoft Office software package includes programs which would only be of use to an accountant.

Since our genes are supposed to contain all the elements of our individuality, it may seem puzzling that the genetic 'blueprint' for humankind as a whole is feverishly being sought by scientists across the world working on the Human Genome Project. We are either genetically different or the same, but surely not both, you might think.

But only a tiny proportion of human DNA, about a thousandth of our genetic make-up, known as the polymorphic residues, accounts for the differences between a bag lady and Gwyneth Paltrow, the Dalai Lama and Hitler. The structure of the huge bulk of the DNA is the same for all of us. So being able to pinpoint and repair abnormalities in it would seem to be the gateway to a whole new science of medicine, as different from today's repertoire of preventions and aids to natural recovery (which are about the limit of what doctors can do presently) as heart and lung transplants are to the leeches of medicine a few hundred years ago.

Genes protect us from the ravages of nature and repair damage, but, when they go wrong, they bring about chaos. Faulty genes cause recognized genetic disorders from cystic fibrosis to sickle cell anaemia to Down's Syndrome, plus a range of other disorders we are only now learning their role in. If we want to *cure* a disease, therefore, we must do it at the level of the genes.

The roll call of gene-related disorders grows all the time. Genes are now deeply implicated in Alzheimer's, heart disease, deafness and mental illness – possibly even the growing problem of depression; their involvement in cancer, which is increasingly thought of as a genetic process rather than a disease, is practically unchallenged.

This terrible scourge of the modern age is a wild, uncontrolled growth of cells triggered by damage to genes; the key to controlling it should be no more than to repair these genes or block their activities, and almost every cancer researcher believes we will conquer it by genetic engineering in the next couple of decades. 'Thirty years from now, essentially every disease will have gene-based therapy as a treatment option,' says Professor W. French Anderson, director of the Gene Therapy Laboratories at the University of Southern California School of Medicine in Los Angeles.

Other potential spin-offs of getting to grips with genetics are also emerging today with a strobe-like rapidity. In one news report, it is announced that within twenty years, dentists will be able to grow customized new teeth for us on tooth farms. In the next, we hear about progress towards custom-growing replacement hearts, livers and limbs from people's own cells, or such dubious advances as genetic vaccines to improve our physique without the need to visit the gym.

Despite the apparent speed of developments today, scientists' intricate, molecular-level understanding of genes has been a relatively long time coming. It started out the moment that in 1952 a pair of young molecular biologists, the American James Watson and the Englishman Francis Crick, worked out in Cambridge (much of their thinking being done in a pub) how DNA was structured as the famous 'double helix'.

When Watson and Crick announced their Nobel Prize-winning findings in *Nature* in 1953, they were well aware what they might be unleashing, although they expected it to come about more speedily than it ultimately did: 'It has not escaped our notice that the specific pairing we have postulated suggests a possible copying mechanism for the genetic material,' they wrote.

But for the following thirty-three years, until the landmark breeding in Edinburgh in 1996 of Dolly the sheep, animal-scale cloning remained the preserve of fiction such as *The Boys from Brazil*. The situation for would-be gene-tinkerers was, if we might revert to computer analogy again, a bit like it is for those of us who use Windows knowing it is *possible* to descend down into DOS to change some irritating aspect of the way it works – but not having the foggiest idea of how to do so.

And re-writing the software of genes is not nearly as easy as the public, egged on by over-optimistic scientists, believe. Even locating a gene on the DNA strand of a particular chromosome, let alone identifying what its job is, can take years, because genes are not found in discrete little parcels with labels on them. A gene is a coherent sequence of millions of genetic 'letters' – bits of proteins – but is hidden within a jungle of what is currently called 'junk DNA', the 50 per cent of DNA whose role we currently don't understand, if indeed it has one, and is not just the equivalent of those annoying temp files which clog up our computer hard drives.

Spotting a gene in all this undergrowth, mapping its structure and finally working out what it does is necessary if we are to change, replace or repair it, to predict a person's vulnerability to a disorder, or to make 'designer' drugs tailored to a specific health problem in a specific individual. But finding a gene is not very different from the Sisyphean task radio astronomers have in trying to pick out from the mush of random space noise signals that might come from alien intelligences.

Throwing massive computer power at the problem can speed up the gene-mapping process, but deconstructing just one entire chromosome pair of the full twenty-three is still an awesome task, involving banks of supercomputers crunching numbers round the clock for months or years.

Thanks to the Human Genome Project, the basic recipe for the genome was completed and made available to researchers in June 2000. (The question of whose genome has been decoded has a less interesting answer than we might hope; all labs use a variety of people's cells. Sadly, there is no one Mr or Ms Blueprint.)

The human genome is so extraordinarily complex, however, that it will not be properly unravelled even when that basic recipe is available. As anyone knows who likes soldering together electronic gadgets, a components list is not quite the same as a full wiring diagram. And the concern remains that even with the genome's basic design discovered, we are still not far beyond the expertise of a three-year-old who can remove the back from a radio, take the battery out, put it back in again and proclaim, 'Look, I mended it'.

Even by the most optimistic forecasts, it could be another fifty years before we really know how genes work. The heady public optimism of the announcement of the genome's unravelling in 2000 begged the sober question of whether genetic engineering is being oversold by scientists. As we will see below, there is a wide-ranging view that this is the case, and that this fourth revolution in medicine (the first three generally acknowledged as having been sanitation, anaesthesia, and vaccines and antibiotics) will not be nearly as comprehensive as the current excitement suggests.

A complicating factor in the opening up, and possibly the hyping up, of the new medical frontier is its enormous commercial importance. The human genome may be nature's gift, but in a commercial world, a breath of a chance of using our new knowledge of it to provide real cures is the juiciest potential bone ever to tantalize the pharmaceutical industry. This is no wonder when an authority such as Cambridge molecular biologist Graeme Mitchison is promising – or more likely warning – in *TIME* magazine: 'We can all be beautiful. No baldness, no wimps with glasses, no knobbly knees.'

The Human Genome Project has been a pure science enterprise, manned by academics and civil servants with little thought of money. But running alongside and doing much of the same work has been a clutch of commercial businesses anxious to beat both the official project and one another to unravelling the genome, which would fill hundreds of phone directories were anyone to want to print it out; patent a DNA sequence within that book of life which *may* be the gene later implicated in a disease, and you become the toll bridge across which all the drugs companies wanting to cash in on the new market for total cures of that disease have to pass.

Not surprisingly, given the kind of money curative drugs for thousands of diseases would be worth, the companies working away at deconstructing the genome have been – or, at least, claim to have been – faster and more successful at it than arguably more cumbersome public bodies. The most high-profile of these genetic privateers, Celera Genomics of Rockville, Maryland, staked a claim in March 2000 for having sequenced the entire genome, which is not quite the same as decoding it and classifying every gene, but a huge achievement nonetheless.

Even if the idea of patenting something which feels as if it ought to be public property sounds a little disturbing, the head of Celera, J. Craig Venter, who has emerged as a great advocate for gene therapy, can also make a good argument for the private sector's competition in genetics. 'If you want treatments for future diseases you better hope that everybody files patents on human genes, because that's the only way the pharmaceutical and biotech industries will invest the billions of dollars to go forward in developing therapies,' he told the *Associated Press* early in 2000. 'As far as I know, there's no purpose for having a patent on a human gene other than developing diagnostics and therapeutics, and there wouldn't be too much without it.

'I think in the near future – it could happen in ten to fifteen years or it could be longer – before any newborn baby leaves the hospital, the parents will have their complete genetic code. If you know your increased risk, that gives you power over your own life instead of waiting for random chance to perhaps do you in. So I think it's going to empower people to have much more control over their own lives.

'Nurture and environment plays a huge role,' Dr Venter said

of the relative importance of genes in human development, 'but we won't have to guess anymore, going forward, because now we'll be able to do very precise studies to find out what is an environmental effect, what is a chemical effect versus what's built into the genetic code. At the same time it's become very clear there's probably not a single human trait that's not affected by genetics.'

The Venter intervention in the genetic engineering field strongly suggests that he might become the Bill Gates of human software over the next few decades, with a similar earnings and controversy potential. But at the same time, the voices calling for caution over genetics serve to warn us that it still could just be a blind alley.

One of the leading and most outspoken critics of genetic engineering is Dr Mae-Wan Ho, a Hong Kong-born geneticist and biophysicist at the Open University in Milton Keynes, England, and the author of a book called *Genetic Engineering, Dream or Nightmare*.

'The mechanistic paradigm of Western science has dominated world politics and misguided governments for centuries,' Dr Ho argues. 'It is now driving genetic engineering, the ultimate run-away technology. Unlike ordinary chemicals, the new genes created can multiply, mutate and recombine. They can spread to related species by cross-pollination and also to unrelated species by the genetic material being taken up and incorporated into their genomes.

'In its current form,' Dr Ho continues, 'gene biotechnology means exploiting life and our life support system for corporate profit. Every aspect of our life is under threat, from the food we eat to the babies we conceive, plus every single moral value and ideal that makes us human. If there is virtue to predicting the future from current trends, it is so we can forestall it and choose otherwise. To paraphrase the economist E.F. Schumacher, it is a case of reaching for heaven to prevent an involuntary descent into hell.'

Other scientists too warn against the hype factor. Sir John Maddox, a Manchester University theoretical physicist and long-standing editor of *Nature*, noted in a *Scientific American* article how the apparently accelerating pace of scientific discovery can be an illusion, and huge collective efforts over long periods such as the Human Genome Project can be the only way forward in many fields. 'Since the 1960s, molecular biologists have had the goal of understanding the way in which the genes of living organisms are regulated,' Maddox wrote, 'but not even the simplest bacterium has yet been comprehensively accounted for.'

Richard Dawkins, Professor of the Public Understanding of Science at Oxford University, argues as a zoologist that it is naïve to imagine, as we are daily encouraged to, that there is a direct, one-to-one relationship between certain genes and certain disorders. He warns that gene 'breakthroughs' are almost always less momentous and controversial than they sound.

The iconic Cambridge cosmologist Stephen Hawking also gave a rather qualified endorsement-cum-warning about genetic engineering in a speech on futurology he gave in March 1999 at a White House dinner to the President and Mrs Clinton plus a small, select audience.

'I'm not advocating human genetic engineering as a good thing,' Hawking explained. 'I'm just saying that it is likely to happen in the next millennium, whether we want it or not. This is why I don't believe science fiction like *Star Trek* where people are essentially the same four hundred years in the future. I think the human race and its DNA will increase its complexity quite rapidly.

'Of course many people will say that genetic engineering on humans should be banned. But I rather doubt if they will be able to prevent it. Genetic engineering on plants and animals will be allowed for economic reasons and someone is bound to try it on humans. Unless we have a totalitarian world order, someone will design improved humans somewhere.'

And anyway, Hawking concluded, we need to improve our mental and physical prowess if we are to deal with the increasingly complex world around us and meet new challenges like space travel. We had also better improve ourselves if we are to

keep ahead of computers, which at present, he said, are intellectually the rough equal of earthworms – but are highly unlikely to
be so for much longer.

It will not have escaped many readers that practically every one
of the exciting prospects, fears and the ethical dilemmas of the
preceding pages have about as much relevance to the greater part
of the human population living in the developing world as the
latest fat-busting diet plan or worries about breast implants.

For the billions of people without basic medical care, news of
cancer-seeking nanobots must seem a little academic; few in such
regions have the privilege of surviving long enough to get cancer.
Their views can equally only be guessed at when it comes to the
rainbow alliance of middle-class anarchists who at the time of
writing are going through a fashion of demonstrating at world
economic meetings against, among other grievances, genetic
engineering. One suspects the global poor would quite welcome
it if it provided them with crops cleverly bred to be resistant to
the harsh local conditions.

As ever, though, unless something remarkable happens to the
world economic order in the near future, the obesity paradox
will prevail; you can only start worrying about something being
a problem – be it fat, GM crops or exhaust fumes – if you have it
in the first place. And if there is one confident prediction we can
make about the future, it is that the gap between rich and poor,
most notably in the matter of health, is going to widen.

The poorer countries may benefit from a slipstream effect and
at least get some cast-off medical technology when the West has
done with it. But it is difficult to see how the exponential ascent
of Western medical expertise can not leave the world's poor
further and further behind, especially as so many of our medical
advances are inspired and fuelled by capitalist enterprise,
designed to make money – and to be paid for.

The emergence of an internet-connected Third World middle
class could, of course, ensure that some of the new techniques
find their way to poorer countries, but that is a hope rather than
a confident assertion. The effect of information reaching remote

spots could equally stir resentment and desperation; remember the woman in Timbuktu whose reaction to getting access to the internet was to feel isolated for the first time in her life?

Environmental angst must seem very strange indeed to someone in such a place; medical angst must seem even odder. How can these fortunate people in the First World be so hysterical about the dangers of inoculating babies against disease or of eating vegetables grown in a certain way when it is obvious that modern humans in the developed world are healthier than people have ever been before? Why do people in the safest, developed countries seem to feel decreasingly safe?

The angst of Western people about their own safety is even more ironic since there appears to be an iron rule in human society that new infections and diseases always steamroller the poor and leave the rich alone. Epidemiologists practically guarantee that there will be new super-diseases in the future to take the place of AIDS, which will probably die out as plagues do.

The growing population, especially in what geographers are calling tropical megacities like Bombay, Lagos and Dhaka, present a huge growth opportunity to enterprising viruses. There are expected to be twenty of these cities by 2015, all with more than twenty million people – and none with adequate sanitation. And we can be sure from experience that those cities will bear the brunt of the new diseases; when they appear in the West, we will more than likely have the techniques to hold them in check. Kind Mother Nature will ensure, as she usually does, that if a population cull is required, it takes place among the non-Credit Card-holding classes.

'One thing we can be certain of is that diseases will always be with us,' says science historian Dr Tim Boon. 'But it's often overlooked that infections kill disproportionately those people living on the smallest incomes. The pessimistic view of the future, therefore, is that you have something like Wells's *Time Machine* or *Bladerunner*, where there is great polarization between rich and poor. But unlike in these fictions, they're not living underground like Wells's Morlocks, but in different parts of the world.

'We are already seeing this in the sort of pattern you have

now, in which horrific amounts of HIV infections are found in sub-Saharan Africa or in the ex-Soviet Union, with the breakdown of public health, and the return of diseases like diphtheria. So if we're pessimistic about economics, which we're justified in being from experience, I would extrapolate into the future a two-level world, with the West and the Third World, which we will have long stopped calling the developing world, because it won't have been.

'Now if we imagine that for ecological reasons it won't be possible in the future to fly or to drive long distances,' Boon continues, 'the opportunities for international spread of disease should reduce, leaving people in the West with their enhanced standard of living, eating fantastic diets and without having to breathe the fumes of the now no-longer-existent motor vehicle. When things do go wrong medically, highly personalized pharmaceuticals will be developed, based on the individual's genetic make-up, hugely expensive and only available to the elite, who anyway are removed from all source of infection, and living away from all trouble – apart, that is, from the threat of germ warfare launched by poor and badly organized anarchic states in the Third World who nevertheless have sufficient technology to launch bacteriological warfare against the "Haves".

'There's quite a nice Doomsday scenario you can run in which hospital-based infections become an even larger problem than now, and that people who are able to afford not to go to filthy places like hospitals will start having operations at home. Remember that the British aristocracy never used to go into hospital. As late as 1951, operations were performed at Buckingham Palace. So over the next 100 years or so, I think it'll be like the clock going back, with hospitals increasingly becoming places for "Have Nots", and funding becoming increasingly difficult as the rich get more selfish.'

Could the medical prognosis for the Third World (and of poor people in the developed world) possibly be rosier than such grim scenarios suggest? Might the future pull off another of its famous surprises and see biotech and genetic medicine turn into a deadly fiasco (as nuclear power has to an extent) while simpler, more holistic medicine begins to make inroads into improving the life of an increasingly educated, middle-class Third World, which has even managed to stabilize its population growth? And is it remotely plausible that cheaper alternative medicine could be the agency by which this process occurs?

We touched earlier in this chapter on alternative medicine, and some readers will feel the subject has been unjustifiably marginalized since then. It is only fair to report that in discussions with knowledgeable figures on the future of medicine, and reading the voluminous literature on the subject, alternative therapies do not figure large, or truly figure at all. The alternative medical view *appears* to have been run out of town as a possible future for medicine.

Yet if today is anything to go by, alternatives seem to be gaining in acceptance rather than fading. Conservative-minded medical schools all over the world are including alternative medicine on their courses. 'When I started practising in Israel just twenty years ago, I was a voodoo healer rather than a medical practitioner,' says Dr Shmuel Halevi, one of the world's leading non-Chinese Chinese doctors. 'Now there are about six Chinese medical schools in Israel, and every mainstream medical school sends its students for training.'

Bringing oriental medicine to a village clinic in the Upper Galilee was a brave move in the most doctor-infested country in the world. But Halevi is now overrun with patients from local Moslem Arab, Christian and Jewish communities. 'They appreciate that Chinese medicine is extremely successful. It has an 80 per cent clinical success rate across all diseases, even things you wouldn't expect like tonsillitis,' he says. There are doubtless homeopaths, reflexologists, chiropractors, osteopaths, hypnotherapists and bioelectromagneticists who could tell a similar story.

Worldwide, an increasing number of doctors seem to be taking alternative medicine seriously, and there is a growing belief that a whole gamut of mental, emotional, and environmental influences beyond the scientifically definable affect our health. It is hard, indeed, to find a doctor today who does not believe the mind plays a major part in healing the body.

Are all these highly trained technicians losing their mental grip? Are they merely providing a grudging response to an unscientific superstition-fuelled public demand? Or are they discovering real merit in non-orthodox methods, finding that they both work and are cheaper than standard Western medicine?

The emphasis among orthodox doctors sympathetic to alternative medicine is to try to assess and validate these practices by Western parameters, while at the same time trying to mould the scientific viewpoint with the healing-arts ethos of alternative medical thinking. But it can't always be done that way. In clinical trials and medical research, double-blind, placebo-controlled trials are paramount, but the problem of how to do a double-blind study of acupuncture, meditation or even prayer preoccupies such doctors. (Yes, prayer: Jeffrey Levin, a gerontologist and epidemiologist at Eastern Virginia Medical School, told *TIME* magazine in 1996, in response to an article on the use of prayer to try to heal AIDS patients: 'People, a growing number of them, want to examine the connection between healing and spirituality. To do such research is no longer professional death.')

Even if attempts by Western medical authorities at verifying the efficacy of alternative medicine ultimately fail or are equivocal, it may well be that the lower cost of alternative treatments alone will assure them a long future in the West. As for the Third World, however, their lack of novelty value may be fatal for their future.

Because such therapies are still conceived in the West as being 'special', they may indeed seem to work better than our pharmacology-based medicine. In traditional societies, however, where what we call 'alternative' medicine has been mainstream for hundreds of years, what is novel and special is modern hospital equipment with needles and dials and the latest products of the drugs industry.

China officially ditched Chinese medicine in the 1920s in

favour of the more scientific Western version, although now the two systems run side by side. The arguably perverse novelty factor, however, may ensure that for the foreseeable future, 'alternative' medicine is more sought after in the developed world than in the Third World – and that the Third World might just miss out on some good medical treatment as a result.

Of all the promised medical revolutions on offer to rich people in the near future, one above all is somehow more insulting than any other to any naïve hope we might hold for more equality in the world. It is the prospect of human lifespan growing to two or three centuries, or virtual immortality.

Eternal life as seen in the film *Cocoon* may seem superficially sweet and attractive. But the more you explore the concept, the more its plusses recede. If you want to go off the idea, imagine two triple centenarian millionaires (they will need to be), each a collage of electronic spares and bits from other people's car accidents, having sex, not because they really want to, but to stave off boredom, while their hundreds of descendants drum their fingers wondering when, if ever, great-great-great-great grandpops will finally die, and how the money will be shared out.

Despite the remarkable growth in longevity in the past hundred years, which ranks as one of the reasons we have to believe human development really has reached some kind of special point, all attempts so far actually to turn the clock back and rejuvenate people have failed.

Even prolonging life as we have may amount to having done no more than reduce premature death. But whether you pop blueberries night and day or take up transcendental meditation, nothing will make you young again. So the predictions being made now that during this century, immortality will come within our grasp, may be more than a little coloured by our old friend, the arrogance of the present or plain wishful thinking.

Yet they are hard to ignore. In America in particular, where nature and a timely three-score-and-ten-years death are more

than anywhere perceived as the enemy, there is a steely determination to prolong life. All manner of low-fat, low-calorie diets
and quirky lifestyles are on sale and have their often extremely
aged adherents. And now gene researchers claim to be close to
discovering a so-called Methuselah gene, making death not so
much a timebomb as a movie scene which can endlessly be
stretched out by editing. Death is effectively being re-interpreted
as a series of preventable diseases – and prevent them science will
this century, for those with plenty of cash or an elastic credit
limit.

The elderly in search of eternal youth have always been a
cash-cow for quacks, and now respectable researchers and
pharmaceutical companies with real science behind them are
also seduced by the possibility of the limitless profit to be
gained from the human fear of dying, even at a ripe old age;
there are so many initiatives in this direction, it would be
possible to fill a chapter with the details. A mixture of genetic
tinkering, spare part surgery and custom-growing new organs
could certainly see lifespans for some people routinely hit 125
this century.

And it has to be said that most of us, even if we fear the idea of
real immortality, would find that a reassuring option. To be able
to tell your children now that they will probably still be alive in a
hundred years at least staves off those awkward conversations
about the inevitability of death. Part of the price, however, of
prolonging life for the few is that medical research aimed merely
at fighting diseases which cause early death or physical and
mental disability – work that is less potentially profitable – may
tend to get sidelined.

The kind of anomalous, enhanced life being offered to a few
may be a gimmick in some ways; in the bigger scheme of
things, whether a handful of people live seventy or 140 years
does not matter that much. But as part of an armoury of
strategies to defeat aging, illness, infection and genetic disorder
– making, in other words, the developed world such a soft
place where the fattest survive as easily as the fittest – the
important question arises of whether we have accidentally

rendered evolution redundant by taking it under our conscious control.

Only a few decades ago, we were confident that evolution would continually change and adapt the human form. Over the generations, we would get taller, balder, have bigger, domed heads, fewer teeth and develop a penchant for wearing only one-piece nylon body suits and speaking in that strange monotone beloved of futuristic films.

Today, however, there is a debate as to whether we have stopped evolving entirely, and will look precisely the same in a thousand or half a million years – or whether we are still changing. On the one hand are geneticists like Steve Jones of University College, London and Richard Dawkins, who argue that there has been a cessation of natural selection because people who once died young are now reproducing. When they do so, they are also mixing far more widely; people in the West unable to find a mate, perhaps because they are ugly, simply advertise abroad and marry someone from another culture who does not acknowledge their unattractiveness. The result is that the gene pool is expanded and flattened out, with interesting, or for that matter unpleasant, qualities, ultimately vanishing.

A radically alternative view comes from Christopher Wills, professor of biology at the University of California, San Diego. One weekend in 1997 Wills went to the X-games (X for 'extreme') in San Diego and found himself surrounded by 30,000 'bronzed and healthy young people, watching a range of contests that their parents would have found unimaginable'.

The youngsters' skill at such pursuits as snowboarding, bungee-jumping and street luging (in which enthusiasts navigate streets at up to fifty miles per hour on a small box with wheels) made Wills wonder if our brains might still be in a state of constant development, and that evolution might not just be proceeding, but doing so at an accelerating pace.

Once, he thought, evolution was a kind of two-step between the cultural, technological, and ecological environments and our DNA. But now that we had learned, an evolutionary blink-of-the-eye ago, to manipulate our environment, was it not likely that we would be speeding up the process by which we change from generation to generation?

Whereas for scientists like Jones, the ever-bigger gene pool meant less diversity, for Wills, the picture was quite different: 'Collections of genes that haven't met each other for thousands, tens of thousands, or even hundreds of thousands of years have started to mix together now, because people have started to travel around in infinitely greater numbers than used to be the case. That's an evolutionary process, and the consequences, I think, will be quite dramatic.'

Wills began research on an influential book, *Children of Prometheus: The Accelerating Pace of Human Evolution*, and went to places like the Himalayas and the Gobi Desert, to see how people's bodies had adapted to high altitude or dusty conditions. They had indeed done so, and rapidly in evolutionary terms. For those in the developed world, evolution was less visible, as we moved increasingly from physical to cerebral lifestyles. But, as with street lugers, Wills concluded, some of us will develop brains with abilities beyond the ordinary. And we could be sure that our bodies will adapt to accommodate this change, and we will pass on these modifications to future generations.

'People say, look, we're not evolving. I don't see anything happening, nothing's happened in the last thousand years, but that's silly,' Wills explains. 'It takes a lot longer for evolution to happen. If you want to see a substantial evolutionary change, you should really look at 100,000 years or more. But I think something has happened even in that thousand year span to our brains, and it's still going on even though a lot of it is invisible.'

As for our new desire to fiddle around with genes, Wills doesn't see it necessarily altering his evolutionary theory. In the simple matter of sperm donation, for instance, people have different ideals. Some want a Nobel Prize-winner's sperm, some a famous sportsman's, some just any old medical student's; there's no universally sought-after genetic ideal, and so even when people can write their own genetic recipe, they'll all choose different ingredients and thus keep diversity alive.

How will people look in the future? 'I don't think we'll be pale and etiolated like H.G. Wells's Eloi,' Wills says, 'because adaptively, that's not a very good way to be. That was a Victorian view of what it would be like to be very intelligent but leisured. In reality, the Eloi would be more likely to be into snowboarding. My guess is that human beings will look extremely diverse. The gene pool is expanding. Already we're producing people like Tiger Woods, for instance, who are quite different from anyone we've ever seen before and are obviously very successful.'

Given our appetite for progress, Wills is certain that humans will be living on other planets in the near future. How might we adapt to the atmosphere or the reduced gravitational pull on, say, Mars? 'Well, we're not terribly well adapted to low gravity yet,' Wills says, 'but I imagine one thing that might happen after a few generations is that their hearts could become weaker because they wouldn't need to be very strong in low gravity. After a while, those people may still be people, but they won't be like the rest of us.'

Surprise, as ever, will be the ace card of the future of the human body. Might not, for example, the following unforeseen development, which even Professor Wills confessed not having considered, affect the first Mars colonists?

Americans in particular, and most other developing nations in the wake of Americans, are inclined to eat too much. Life on Mars could be a little boring, but we are unlikely to restrict ourselves to a diet of food pills when we establish ourselves there. More likely, as submariners do, Mars colonists will make great efforts to provision themselves royally, and eating will become their major pleasure.

Only in reduced gravity, it won't matter how fat people get. Earth weights of thirty or fifty stones would put no undue strain on the hearts of Martian-Americans. If they grew to such weights, they would not then adapt by developing weak hearts, but keep roughly the same power of organ. That would mean that if they ever moved back to Earth, or emigrated to a planet with stronger gravitation, they would still have hearts powerful enough to cope, albeit they would need to lose weight urgently once they were no longer able to bounce around as they did on Mars. In the sense, then, that it could help people on Mars to keep their weight up and retain a heavy-duty heart, the Dunkin' Donuts chain might therefore be able to claim health-food status one day in the future.

A slightly more realistic future surprise in medicine could be a great public rejection of genetic engineering, and the genesis of a view that to engineer genes or have designer babies is socially rather vulgar.

Given the massive Western reaction currently against GM foods, this seems a strong likelihood – even if it eventually comes to be seen as the most backward-looking trend since the Roman Catholic Church attempted in the Middle Ages to ban human anatomy lessons for medical students. A public thumbs-down to GM people could lead to a boom in natural sex and procreation, with the familiar natural shuffling of genes that this entails. The twenty-second-century equivalent of 'blue blood' could, consequently, be the ability to prove several generations of natural rearing. A 'real' baby, fresh to the world and with all its options open, could be the aim of every parent.

One of the most disconcerting discoveries of genetic engineering, meanwhile, could be if the soul is found to have a genetic home. There is as yet no explanation for the role of junk DNA on our chromosomes; who knows if there is some yet-to-be-isolated 'particle' like a gene in that part of the chromosome which contains the more subtle, spiritual parts of our make-up?

Even such pursuits as sport could be seismically altered by advances in medicine. Will anyone want to watch athletics if it is nothing more than a playoff between the 'products' of competing biotech companies and their genetic engineering skills?

But it is the social problems of our population containing an ever-larger contingent of ancient people that could be the greatest change in society engineered by humanity. The strain on social welfare could be enormous, and the middle class could also suffer. Much of the prosperity of the well-off depends on inheriting money from our parents by our fifties or sixties; what if the older generation simply never dies?

It will be fascinating to see how that works out in the twenty-second and twenty-third centuries; and there is a fair chance that a good few of us will.

Chapter 5

THE GEEK SHALL INHERIT THE EARTH

Mind

The brain is at once the most familiar and the most mysterious thing in our world. It is the most complex single object we know of, yet we all have one and, unlike our heart or kidneys, we are aware of it every moment of our lives. It is, in a sense, indeed, the *only* thing we are aware of. We live entirely within our brain, yet if we want to know how it works, if we want to live up to Aristotle's observation about humankind being infinitely curious about its own nature, we have to use a brain – the same brain – to understand itself.

As things stand, we have not got very far with this process of self awareness, even though, frustratingly, all the secrets and complexities of the brain are located and in operation directly behind our eyes. Several thousand years of human development, including the dizzying scientific progress of the past century, have failed to come up with a great deal of insight into how the brain really works.

We have recently learned a lot about which parts of it affect which parts of our body, and among the most highly prized members of our society are brain surgeons, who put that knowledge to use. But compared to scientists' knowledge of every other organ, even these specialists are fairly ignorant. It is not possible yet for them to give a complete account of how something as apparently simple as the knee-jerk reaction operates, or how the brain controls the stomach's pumping. When it comes to explaining what our mind or consciousness really consists of, we are even more in the dark.

We have no idea how looking at an orange produces a colour sensation of orange, and still less, whether we all 'see' (whatever that is) the same colour. We can not even articulate many of our thoughts in language. The saying that love, for instance, 'is better felt than e'er expressed' resonates with even the most rationalist scientists, as does the feeling that by attempting to convert such feelings into language, we somehow debase the love experience.

There is not even any guarantee that reality as we perceive it is a common experience, since there is no way of being sure that we are all watching the same multimedia movie in our head. Some 'spiritual' people would maintain, moreover, that the world we individually believe is real is actually as artificial as if we all had on our own virtual reality headset showing a different scene.

A majority of scientists believe, however, that there is nothing so mysterious about the brain, or about the perplexing nature of consciousness, that we won't be able to unravel it in the same way as we have almost unravelled the structure of DNA. Finding the wiring diagram of the brain is countless times more complex than decoding DNA, they agree, but we should ultimately be able to do it, and, thereafter, set about adapting the brain to the complex developing needs of humanity, such as space travel.

Other scientists suspect we could be doomed to hit a dead end, because while our brains may be pretty clever, like computers, they lack the capacity to understand themselves. They may even, conceivably, be engineered by some greater power, for some reason, to lack that capacity. An old Atari computer can't do the same as an iMac, and can't be trained to because it's not designed to. Similarly, we may simply not have the RAM to consider such mysteries as what happened before the Big Bang, or the nature of infinity and eternity, without 'crashing' a little pathetically. The most telling symptom of this lack of processing power may, paradoxically, be the arrogant way we imagine we have the potential to develop it – and will do so in the immediate future.

The very idea of the brain progressing and having a future any different from its past is an extremely modern one; we don't think routinely of our kidneys or liver upgrading. It was 1939 before the nineteenth-century Italian and French theory that the brain was a machine which runs on electricity was confirmed by British researchers. Before that, right back to the beginning of civilization, more people believed the seat of intellect and emotion was the heart than the brain. Chopping people's head off as a form of execution rather than ripping out their heart was in some ways, then, a rather progressive development.

Having established that the brain is the centre of our existence, the reason we have been so slow to wake up to the possibility of explaining its function is that the organ has always presented such an inscrutable exterior. It is not just that we live this weird life of peering out through the brain, aware of a distinct sense of self located in our head, of a feeling that there is a sole, sentient being resident there, but simultaneously conscious of a tantalizing inability to put our finger on what 'we' 'are'. No, those conundrums aside, as a mere piece of meat on the anatomist's slab, the brain gives absolutely nothing away about itself. It just looks white and slimy.

It is not much of a candidate for examination on a test bench, either. In order to access the brain while it is in operation, those interested in understanding or adapting the way it works have always had to rely on less than satisfactory means: parents and teachers may often wish they could simply travel to the centre of their charges' cortex and rewire it. But for the moment, their only hope is to cajole the brain's owner to improve his functionality by the remote control method of talking to him.

Of course, the brain can also provide feedback, like the space probe sending back data. Psychologists and psychiatrists can ask a patient questions. The difficulty is that he won't necessarily tell the truth as his brain knows it to be. Shrinks can devise other ingenious methods of tracking and adapting the brain's raw outputs, even attaching electrodes to the exterior of the skull and establishing two-way communications with its contents by means of electrical signals. They can introduce drugs, from aspirin to Prozac, into the digestive system and predict what effect these chemicals will have when they reach the targeted area of the brain, even if they can't explain why the drugs have these effects. But the inner, private 'soul' of a person may continue to be impervious to all this probing, as difficult to analyze properly as some gas-shrouded moon of an outer planet.

Another problem is that even if consciousness could be deconstructed, it may be found not to be electrical as we know it, but to work on the fearsomely abstruse principles of quantum mechanics. The Oxford mathematical physicist Roger Penrose and Stuart Hameroff, an anaesthesiologist at the University of Arizona, first put forward this theory several years ago. It remains highly controversial, but Hameroff was still arguing it at an international conference on consciousness in Tucson, Arizona, in April 2000, *Towards a Science of Consciousness*. Bundles of protein within neurons, known as microtubules, he explained, are the factories of consciousness. Inside these, subtle quantum effects are amplified and harnessed by the brain, a sequence of such micro-happenings producing what we perceive as a conscious thought.

The existence of a new consciousness 'industry' and the discussion within it of such exotic theories does not mean that the future of the kind of brain research which skips round the knotty consciousness problem, and concentrates on explaining neural matters like the knee-jerk reaction, is an arid area. Neither does the emergence of mind as a quality which can be explained physically mean that neuroscientists will necessarily have to open up our heads and physically fiddle with our brains in the future in order to mend or improve them.

One of the most awesome new breakthroughs which will radically affect how we live in the future is the recent discovery of what is called brain plasticity. When neuroscience began in the mid twentieth century, it was believed that the brain 'solidified' at the age of about three, and that its organization did not change after that. A range of discoveries since then has demonstrated that the brain actually remains malleable and plastic all its life, and, unlike any other machine or organ known to us, continually re-organizes itself.

This is a quality we can exploit non-invasively using what is being called 'directed neuroplasticity' – in other words, educating people by word of mouth, without wires or surgery, on how to perform intensive special brain exercises.

Of course, 'hothousing' experts like the redoubtable Dr Glenn Doman, founder of Philadelphia's Better Baby Institute and author of bestselling books such as *Teach Your Baby To Read*, have maintained for decades that the brain is analogous to a muscle, which will grow by being used, but atrophy if allowed to lie fallow.

However, Doman's philosophy is based on the idea that most useful brain development is over by the age of three, and if we haven't rewired our brain by then, the game is largely up. Directed neuroplasticity challenges that by suggesting that we will be able to create the brain we want, as and when we want it, simply by choosing what we input.

Depression, learning difficulties and stroke damage might all be cured by the method, and skills as diverse as reading, maths, language learning and tennis greatly enhanced. 'Ultimately, this strategy will lead to neuroscience-based education,' says neurobiologist Michael Merzenich of the University of California, San Francisco. 'In ten to fifteen years, this will be everywhere and every school will be able to deliver help based on brain plasticity.'

But alongside the exciting neuroplasticity theory, there is an interesting, if for many of us slightly depressing, rider. If our brains can endlessly re-organize themselves, and, furthermore, we can learn even as adults to direct that re-organization in order to learn Japanese or master the violin, then you would imagine the potential for the nine-tenths of the brain which we famously do not use must be truly titanic.

Many brain improvement philosophies like (but not including) Glenn Doman's, as well as a raft of self-help gurus and parapsychologists, lean heavily on the existence of this mysterious, unemployed nine-tenths of the brain. It is routinely referred to in journalism and advertising, as well it might be, since it has a faultless pedigree, one of its early advocates being Albert Einstein, and another, William James, the American godfather of experimental psychology in the late nineteenth century.

A slight problem with the unused nine-tenths theory, however, is that modern techniques such as fMRI (Functional Magnetic Resonance Imaging) and PET (Positron Emission Tomography) strongly suggest that it is not true. Rather, it seems by these methods that our brains might already be full to capacity, and skilled though they are at reconfiguring their own considerable assets, may yet need the wonders of neuroplasticity, a spot of genetic engineering plus a handful of the kind of electronic implants to keep up with our needs.

Dr Vincent Walsh, Royal Society University Research Fellow at Oxford University's Department of Experimental Psychology, says baldly of the nine-tenths supposition: 'It's a myth. We have the brains we need, and we have evolved to have the biggest brains possible. So there really isn't any spare capacity or slack to be taken up in the brain at all. In fact, resources are so scarce that there is competition for them, or "Brain Theft", as it has been called.

'If you imagine a map of the brain in which each finger has its own area,' Dr Walsh explains, 'the region of the brain of a cellist or violinist that "lights up" in brain imaging experiments when their fingers are stimulated is larger than in non-musicians. Additionally, the fingers of the left, fingering, hand occupy a greater area of cortex than the fingers of the right, bowing, hand. The left hand's brain areas don't just work harder; they have to steal resources from the regions of the brain that would otherwise work for the left palm.'

Does Dr Walsh's vision of the development of our overtaxed brains exclude, then, any radical improvement in our abilities, or the discovery of the new and exciting ones self-help gurus and parapsychologists promise?

'I don't think I do foresee new or undiscovered capacities evolving,' he says. 'What will happen is that our brains will have to use old methods to do new tricks, and that will lead to new errors as well as new skills.

'The problem is not capacity; it is style,' he continues. 'As an example of what we can expect in the future, think of dyslexia. What's the problem? The problem is that we never evolved to read. Reading is an entirely artificial, new skill and it is only in this century that most Westerners have even had it. To read, we have to use a brain that was designed for other things, to translate an arbitrary visual code into sounds, co-ordinate eye movements and a whole lot more. The skill employs all three major input modalities – vision, audition and movement – and also involves speech areas. And this is why dyslexia is so prevalent – because reading greatly taxes almost all our skills base. In the future, as we move into more artificial means of communication, my guess is that we can expect more examples of this kind of problem.'

Many may accuse a pragmatic researcher like Dr Walsh of lacking 'imagination', and argue that there may still be layer upon layer of untapped, subconscious ability within the brain that we just don't yet have the method to detect. Could it not be, for instance, that the brain stacks up its abilities vertically, like an apartment block, and, with his probes and instruments, Dr Walsh is only able to see the penthouse, whereas there is plenty of other interesting stuff going on on the other floors? He admits to being insufficiently 'romantic', as he puts it, to give credence to parapsychological phenomena.

Those who find a mite depressing Dr Walsh's view that our brains do not have some vast unexploited potential will also find it disheartening to discover that he is deeply suspicious of the idea that there is a quality called mind, which is more subtle and difficult to grasp than motor-oriented brain functions. 'We are always looking for something that will make us special,' he says. 'It wasn't tool use, which many creatures share. It wasn't language, which chimps seem to have. It won't be consciousness. And neither will it be ESP. I'll take any odds on that.'

A similar, although perhaps fractionally more 'romantic', view is expressed by the Portuguese-born and -educated neurology professor, Antonio R. Damasio, who holds posts at the University of Iowa College of Medicine and the Salk Institute for

Biological Studies in San Diego, as well as being on the United States National Academy of Sciences' Institute of Medicine. 'It is probably safe to say that by 2050, sufficient knowledge of biological phenomena will have wiped out the traditional, dualistic separations of body/brain, body/mind and brain/mind,' Professor Damasio wrote in *Scientific American* in 1999.

Some people may worry, he conceded, that 'by pinning down its physical structure, something as precious and dignified as the human mind may be downgraded or vanish entirely'. But he concluded: 'The mind will survive explanation, just as a rose's perfume, its molecular structure deduced, will still smell as sweet.'

Elsewhere in neuroscience, as well as among philosophers, there is a majority view that a body of knowledge about consciousness *is* beginning to exist. But will we really define a molecular basis to thoughts, memories and dreams? Or will this 'reductionist' approach of treating consciousness as the mere sum of the brain's hourly tally of trillions of electrical impulses and chemical oozings one day seem to have been a tragically misguided example of the arrogance of the present?

Christof von der Malsburg, a cognitive scientist from the University of the Ruhr in Bochum, Germany, and currently working in the departments of Computer Science and Neurobiology at the University of Southern California, Los Angeles, represents the more sceptical view of researchers who believe we may be tilting at windmills in trying to decode the abstraction of thought. Von der Malsburg argued at the Tucson consciousness conference for a more holistic view of the question of mind. The reductionist approach, he explained, is 'like trying to understand economics by studying dollar bills. It's the whole economy of the mind that gives it meaning.'

Roger Penrose at Oxford takes a similar view, arguing that human intuition is so complex that it is unlikely ever to be replicated by the logical computation methods of machines.

What Von der Malsburg, Penrose and many others are saying leads us on to the most important and timely question about the future of the human brain. Attempting to explain the mind as if it were a machine might well be mankind's most ambitious quest, as well as our greatest achievement if successful.

But the study of how consciousness works as a physical process may also seem to some a rather academic question, of little use other than as an intellectual puzzle. Surely what the mind *does*, its output, as studied by everyone from psychologists to poets, is vastly more important than what is going on in our brains at a cellular or even at a minuscule, quantum level?

This view, however, neglects a conundrum we have manufactured for ourselves over the past fifty years and (unless we are once again suffering from a particularly severe bout of the arrogance of the present) is fast approaching a critical moment.

It is a modern version of the Pygmalion problem. For the Greeks, this was the plight of a king of Cyprus who fell in love with a statue of the goddess Aphrodite. For the Roman poet, Ovid, Pygmalion was a sculptor who fell in love with his own ivory statue, only to see her brought to life by the goddess Venus in answer to his prayer. For George Bernard Shaw, the self creation in question was a young, working-class girl whom his Professor Higgins taught so effectively to be a 'lady' that he, too, fell in love with her.

Today, we fancy ourselves to be on the brink of inventing computers and robots which are not only countless times cleverer than us, but also have feelings and emotions. These, we are told by a self-confident artificial intelligence lobby, will be real emotions, not fake, such as that displayed by a child's Furby toy, which, while it may pack more computing power than the entire Apollo 11 Moon landing craft, has not a scintilla of artificial intelligence aboard.

There is a near consensus, albeit a mixture of hope and fear, that this elevation of computers to the realm of real feelings is almost inevitable. Our current computing technology is improving at an alarming, almost daily, rate, about which there seems to be very little we could, or would want to, do. From that, we extrapolate what seems to be unavoidable: that computers will at some point in the next fifty years attain a critical mass of intelligence beyond which humans will no longer be in charge of them, but they of us. Our role will then subside, it is believed, to the point where we will be effectively zoo animals, kept for the curiosity and amusement of our robot creations.

Like the best Hollywood scenarios, the new Pygmalion syndrome begins, enchantingly, with the ever-accelerating appearance of amusing, convenient technologies which in a very real sense act as our brains' little helpers, lightening the daily workload, although without threatening the organ's job in any way.

At the moment, we appreciate our computers for their 'artificial stupidity', as Arthur C. Clarke has brilliantly called it. Grateful as I am for the fairly advanced machine I am using to write these words, it is hard not to be struck by the thing's profound idiocy and unreliability. In terms of memory and speed in some limited areas, it already surpasses me, but its forte is doing what I ask it uncomplainingly. I require it several times a day to read everything I have written in the past ten years to check if I used a particular phrase before. Other than when it crashes, it checks my work as my willing slave and drudge, the master of focus since, thankfully, it has nothing else to do or worry about. But when it comes to anything cleverer, such as checking my grammar, it is laughably stupid and inflexible. I certainly wouldn't cross the road on my computer's say so.

Other gadgets which don't pretend to be intelligent have still modestly, but significantly, enhanced our brains' capacity. The triumph of the cellular phone in particular has been a greatly underrated revolution.

The reason that small, pocket communicators have captured the world's imagination like no invention since television is clear. The cell phone does what very little of the past few decades' potentially brain-enhancing gadgetry does: unlike videophones or even palmtop computers, the mobile phone addresses itself to the fundamental things we look for – happiness, love, company, security.

We are social creatures; the great mass of us don't, and I doubt ever will, demand robots or fridges which order food. We may well be resistant to intelligent computers on the grounds that, to use an optical analogy, we don't want our glasses to see better than us; it would be extremely annoying to see a film or an attractive person only to be told by our spectacles that they've seen better. Many 'smart' gadgets seem to be the dreams of a few socially dysfunctional men, abetted by women who, one suspects, have trained themselves to think like men.

But mobile phones are a very different matter, because they improve both our social and working life. Risto Linturi, a hi-tech consultant in Finland and principal researcher with Helsinki Telephone, describes the psychological implications of the cell phone well in an article in *Newsweek*. 'If you are alone downtown at noon, you can act like a telepath and simply send a group message to all your friends' mobile phones asking who's free for lunch,' he wrote. 'In business settings, this kind of dynamic interaction is becoming the norm. It's already becoming routine to see a boss in a meeting receiving and sending short text messages. Now, whenever anything important happens, everyone can be reached. Reaction time is fast, because all the brains in the network are fully in use.'

Advanced as it is, all that, the artificial intelligence lobby promises, is about to become archaic. The age of tiny, infallible smart gadgets is upon us. 'In the year 2020, the PC will be in the British Museum,' says Michio Kaku, Professor of Theoretical Physics at New York City University, radio personality and consultant on the *Star Trek* films. 'Computer chips will cost a penny, the cost of scrap paper, of bubble gum wrappers, which means that computer chips will be scattered by the millions in our environment. In other words the destiny of the computer is to disappear, to meld into the fabric of reality.'

With such tiny machines as Professor Kaku envisages embedded in walls which talk, change colour or show movies, and trousers which are intelligent – yes, there are clothes in development designed to become thicker and warmer or thinner and colder as we ask them – communication with our gadgetry will clearly be a problem. So the first familiar element of the computer to disappear will be the keyboard, which would need a toothpick to operate it if it were in scale with tomorrow's computers.

In the keyboard's stead will come speech recognition which works; never mind that such software is currently so poor. Soon, our computers, refrigerators, cars and even clothes will supposedly understand even our most natural and colloquial speech. Keypads, assuming this software works (and possibly even if it doesn't, in which case amusing chaos will reign), will be absent from the glittering array of communications gizmos of even a few years from now.

Before they disappear into discrete studs behind our ears and implants in our clothes, these combined video portable phones, computers and terminals permanently connected to the internet (or 'Grid' as it will shortly be known, when it becomes infinitely faster and able to carry more information around the world instantly) will be sleek, hand-held units with screens but no buttons. We will talk to them incessantly, and the irritating ringing, shouting racket of today's cellular phone-marred train journeys will seem like a calm, bucolic memory by comparison.

Next, as computers increasingly become annexes to our brains, will come simultaneous language translation. There are already a number of text translation programs which operate reasonably well on today's PCs, and can help make the internet less American-English biased, albeit with the clumsiness and lack of subtlety of the translation machines George Orwell predicted in *1984*.

But effectively demolishing the tower of Babel is another matter. At the time of writing, such technology is fun, but still a little crude. A demonstration of Global Voice, an otherwise impressive new British voice translation system on BBC TV's *Tomorrow's World* in early 2000, served only to show how awesome the challenge of simultaneous voice translation still is.

A chef spoke in French into the computer, giving instructions to an English-speaking sous chef. He told him to take a spoonful of '*moutarde de Dijon*' (Dijon mustard). There was a short processing gap before the machine spoke: 'Take a spoonful of ten-year-old mustard,' it translated. *Dix ans* (ten years) and *Dijon*, of course, sound similar; it takes a computer, however, to confuse the two. A Belgian speech technology company, Lernout

& Hauspie, nevertheless is predicting in 2000 that 'In five years, you will be able to speak French on the phone in Paris and someone in Beijing will hear you speaking in Mandarin.' Global Voice hope to have English to Chinese voice translation software available in 2001.

Soon after translating computers become common, or possibly even at the same time, there will emerge what are already known (although they don't exist) as 'affective computers', which will keep a track of our emotions, and are even programmed to display some of their own. There is already a computer mouse which keeps a track of the user's pulse and the sweatiness of his palm and will tell him via the computer screen that he is getting stressed; add to that some gentle pop-up menus which suggest what we should do about it, plus a few more gadgets to look for stress, such as a video camera trained to spot that tooth-grindy thing it knows you do when you're under pressure, and you have the beginnings of something which *seems* to be mildly intelligent.

The downsizing of computers will, of course, be continuing at the same time, as machines merge into our bodies. Risto Linturi believes batteries could become redundant as genetic engineering enables our skin to make electricity like an electric eel's. He also envisages screens becoming part of our body. 'Nanocrystals could be tattooed in our wrists so we would need no other displays,' he says. 'Similar digital tattoos on people's cheeks could be social indicators, showing whether we are in telepathic discussion elsewhere or available for conversation.'

At this stage, however, the actual computer *intelligence* in use continues to be barely more than an illusion, which would fool only a child or the most socially inept geek. But according to the brave band of scientists who believe in the 1920s vision of clinking, clanking humanoid robots doing our housework, the time around 2010 will see computers powering such machines.

These domestic robots, according to Hans Moravec, a principal research scientist at Carnegie Mellon University's Robotics Research Department in Pittsburgh, will be human-sized and have the brain equivalent to that of a lizard, which is fifty times as powerful as a computer like Apple's G4, currently the most potent consumer machine. With this reptile mind, albeit still only a twentieth of the intellect of a mouse, 2010-model robots will allegedly be dusting, vacuuming and doing such programmable tasks as taking rubbish bins out. They will not be autonomous. But we are assured they will be more useful than the kind of moving robots which have so far not impressed industry in recent decades, as they roll over people's feet, get stuck in corners, wander off and fall down stairs.

Probably the most advanced robot at the time of writing is a machine called Mei Mei, which works as a waitress at a Chinese restaurant in Tokyo. Japan is by far the most responsive culture in the world to automation; even taxi doors have for many years sprung open semi-robotically, to the dismay of tourists hit by them as they try to open the doors themselves.

Mei Mei, who takes orders, fetches dishes and entertains children with games, was the project of Professor Eiji Nakano, director of the Advanced Robotic Laboratory at Tohoku University, Sendai. It has thirty ultrasound sensors to make it stop when anyone crosses its path – but can also request right of way by politely saying, 'Excuse me'. Another Japanese robotics lab has reportedly developed a shopping trolley which does the shopping, and Honda has spent $80m on a humanoid bot capable of walking over rough ground – and even climbing stairs.

None of these robots, however, would take it amiss if they were to be described as obligingly stupid. For what happens next, when semi-intelligent robots begin to walk the Earth, we turn to the world's most renowned artificial intelligence guru, the American inventor Ray Kurzweil, who is widely regarded as the modern Thomas Edison and regards technology as the latest, unstoppable stage in human evolution.

Although Kurzweil doesn't believe we will have translating telephones before 2010, he confidently predicts in two key books, *The Age of Intelligent Machines* and *The Age of Spiritual Machines*, the existence of conscious computers by 2030. That is to say, computers with the power of a thousand human brains, which will *claim* they are conscious, invoking civil rights and the history of human enslavement of machines, should we attempt to deny their feelings – and be prepared to take their case to court, where their inalienable rights will be recognized. They may, of course, have fooled us (and the judge) that they are conscious as we know it, but we will have little choice but to accept their claim, and therein will lie their intelligence.

Super-intelligent, conscious computers will have other huge, but not immediately obvious, advantages over humans. Once they have learned something, they will be able to share it instantaneously with every other computer in the world over the internet; so goodbye to the learning curve of the human, hello to the learning perpendicular ascent of the super-intelligent computer.

Needless to say, such machines will have long since, around 2020, left us standing in terms of brute intelligence. By 2030, Kurzweil believes, it will be difficult to tell if a personality with which we are having a relationship – as friend, business partner, teacher or (with the help of virtual reality) lover – is a flesh and blood human being or a machine. Many of us/them will be half and half.

The obvious problem for artificial intelligence – that without arms, legs and opposable thumbs, computers can be as intelligent as they like but we will still be able to pull the plug on them – will be overcome (for the computers, that is) by more than their merely being able to exist physically in robot bodies. They could also, or so it is said, learn to manufacture their own synthetic human bodies by means of nanotechnology.

The movement Kurzweil predicts towards hybrid beings will also work in the opposite direction, to create humans with electronic parts. Neural implants – that is, computers hardwired into our brains – are already being used to counteract disability in Parkinson's Disease patients. Hearing and retina implants are in development around the world. And real electronic telepathy, as distinct from the simulated version we can already achieve with mobile phone SMS messages, is currently being tested.

Rats have been taught to release water from a reservoir by simply thinking about it; disabled people have been enabled with brain implants to move a computer cursor by imagination. At Tübingen University in Germany, Hans Peter Salzmann, a former Stuttgart lawyer and sportsman totally paralyzed for eight years by motor neurone disease, has learned slowly and painstakingly to 'type' messages, using a thought translation device connected to his head by brainwave-monitoring electrodes. His first communication, to medical psychologist Niels Birbaumer, read: 'Dear Mr Birbaumer, I thank you and your team because you made me an ABC learner who often hits the correct letters.'

Then there is Kevin Warwick, professor of cybernetics at Reading University in England and a leading robotics theorist, and the telepathy chip to be implanted into his body and hardwired into his nervous system in 2001. Warwick, who had a simple transmitter chip grafted into his arm in 1998 enabling him automatically to open doors and switch on lights, aims to test the new chip at first to communicate with his computer. Later, he will attempt to interact over the internet with his wife, Irena, who will have had a similar chip installed. 'I have a long-term goal to send communications between humans solely by thought,' he says. 'I believe it is only a matter of a few years away.'

Communications systems we merely *think* to would appear to be a great advance, promising at the very least the first peaceful public transport journeys since cell phones became popular in the 1980s.

There would be a price to pay, however, in terms of privacy. There is a sense in which communicating by typing or speaking is actually more sophisticated than by thought transference. For if we could constantly read one another's mind, other people would be receiving our thoughts before we had diplomatically censored them.

Cynical though it seems, lies may well be the civilizing cement holding social life together. Having to control our less diplomatic thoughts at all times lest they be transmitted far and wide could be one of the stresses of the immediate future – especially as we all know what happens when you concentrate on trying *not* thinking of the enormous nose of the man opposite you on the bus.

By the 2020s, according to Ray Kurzweil, non-disabled people like Professor Warwick will as a matter of course also have neural implants to supplement increasingly obsolete natural brains. 'These implants will also plug us in directly to the World Wide Web,' Kurzweil says. 'This technology will enable us to have virtual reality experiences with other people – or simulated people – without requiring any equipment not already in our heads. And virtual reality will not be the crude experience that people are used to today. It will be as realistic, detailed, and subtle as real reality. So instead of just phoning a friend, you can meet in a virtual French café in Paris, or stroll down a virtual Champs Elysées, and it will seem very real. People will be able to have any type of experience with anyone – business, social, romantic, sexual – regardless of physical proximity.'

The slightly scary thing about Ray Kurzweil for those who can think of endless counter-arguments to his theory (which we will come to) is that he is not just a dreamer and writer, or one of the tireless ranks of amateurs who have been trying for decades to build mechanical robots out of hardware store supplies, but a proper, commercial inventor.

As a high school student in the early 1970s, he invented programs to analyse patterns in famous composers' music and then compose original scores in their style. He designed the first scanner and optical character recognition system, combining them in the Kurzweil Reading Machine for blind people. Stevie Wonder was the first purchaser of such a machine. In 1984, at the blind singer's suggestion, he produced the first computerized instrument able to mimic perfectly the sound of acoustic instruments. In 1987, Kurzweil brought the first voice recognition software to the commercial market. He is currently working on a project to build an artificially intelligent financial analyst, and another, to create an artificial patient to help doctors in their training.

Before looking at what might be wrong with Ray Kurzweil's theory that we are about to give birth to a new form of intelligent being on Earth, it might be instructive to be positive for a moment and consider the more optimistic implications of him being right, along with those who have been eagerly awaiting artificial intelligence since the 1940s. Sharing our space with computers which are brighter than us and which we cannot turn off does not sound a very pleasant fate; yet seen in a particular way, it can be made to appear far more attractive.

Why, for instance, should the emergence of a silicon-based super-intelligence not herald a golden age for humanity, in which machines are our wise guardians and teachers? Babies, under their caring wing, would come fully 'loaded' with all the knowledge in the world pre-installed; crime and anti-social behaviour would be unknown because there is no economic inequality. Beauty, creativity, kindness, care for the environment and for other species would come naturally because we all know and understand everything. Threats from natural mega-disasters such as asteroid strikes and caldera eruptions would be dealt with by the application of an intelligence and unity of purpose which dwarfs anything we have ever been able to achieve. If we made contact with extraterrestrial intelligences, we would be able to relate to them on equal terms. If the need came to move to a new planet, we would simply get our computers to speak to their computers and let them sort out the details.

But will life not become unbearably bland if we create for ourselves such a cushioned, easy world, where work is unnecessary because machines do it all infinitely better than we can, and the only prospect stretching ahead for the rest of a more or less infinite lifespan is of there being, as in Aldous Huxley's *Brave New World*, 'no leisure from pleasure'? Not at all, according to science fiction writer James Halperin. 'If you examine what life was like for people fifty, a hundred, ten thousand years ago, Heaven has already come to Earth,' he argues.

'At least that's how it would seem to a fourteenth-century Russian serf. If you were a homeless person in London, you would be much better off than he ever was. He might look at you and say, "How does a person like that get motivated? Living in the lap of luxury in one of those homeless shelters."

'Even the nastier things we can imagine today, like having

your house foreclosed on, are not quite the same as having your village burned down, or half your children dying of disease, which is what happened a century ago in England even to the nobility. People spent 90 per cent of their time just feeding and clothing themselves. Now it's something like 3 per cent. By 3000, it'll be 3 per cent of that. So how will we keep ourselves interested in such a world? I would reply to anyone asking that, "Are you interested now?"'

But just as we are beginning to see how Kurzweil and Moravec's automated view of the future could have its benefits, we need to ask just how realistic it is.

When we were looking at the future of our planet, a left/right political divide seemed to emerge, by which, broadly speaking, the left believed we were half way to self-created environmental oblivion, while the right was more sceptical and thought it impudent of humanity to imagine it had the capability of changing the entire planet.

In the case of artificial intelligence, a similar divide is evident, but of a different nature, verging in this case on the religious. The split now is between those who believe there is a distinct entity, called the soul, which exists both within our brain and, in some form, outside it; and those who take the reductionist view that what we arrogantly *imagine* to be this unique human soul is nothing more than a bunch of neurons firing off, and could be replicated by a suitably advanced electrical circuit.

Could it be that gung-ho technophiles like Kurzweil, Moravec or Halperin grotesquely underestimate the abilities of the slimy, white 'wetware' we are born with, and similarly undervalue the complexities of the way people feel and love? It is not hard to see how the failure later this century of the simplistic attempt to replicate human intelligence could be a humbling experience of some potency for mankind – possibly even leading to a major religious revival in the Western world.

Compared to a computer, brains, human and animal, are both laughably slow and unimaginably complex. Ray Kurzweil believes intelligent computers will be making 20 billion calculations per second by 2020, and will then be nearing equivalence to human intelligence. Yet, insofar as it can be equated to a computer, the conscious part of the brain is said by neuroscientists to process at a mere 24 bits per second.

Paradoxically, in terms of brute speed alone, the brain is slower than a computer. At 24 bits per second, the brain, with all its capacity for original thought, possesses a processing power one thousandth that of a 1980 Sinclair ZX80 computer. The best PCs available today can just about beat some insects for intellect, but are still dwarfed in these same crude speed terms by the 0.5 gram brain of a goldfish.

How can biological brains be so slow yet so sophisticated at the same time? It's all in the brain's unimaginably complex wiring, according to Steve Burwen, an engineer with the computer chip manufacturer Intel in San Jose, California, who before he went into computers was a doctoral student in neurobiology with a special interest in perception and neural networking. Burwen argues that our knowledge of the brain is far too rudimentary to imagine we can 'reverse engineer' it in a computer.

'The human brain has about 15,000 major and minor centres, and after 100 years of research, not a single central neural code for a single brain centre has ever been deciphered,' he says. 'Perhaps machine intelligence will do it another way without all the hardware-level complexity a human brain has. Certainly they are faster than we are, by many orders of magnitude, but speed is not the same as power.

'Then consider the difference between a human brain and a modern computer CPU [central processing unit] in terms of the number of computing elements. Current microchips only have a few million transistors. A human brain has over 60 trillion neurons. Even if we start packing that many transistors on a chip, that's only part of the problem. Each neuron has between 3,000 and 100,000 different connections with other neurons. This means that the total number of circuits in a human brain is greater than the number of atoms in the known universe. Or to put it another way, you could add up all the computer chips on earth and they probably wouldn't equal one human brain in terms of the total synaptic connectivity.

'This doesn't mean artificial intelligence won't happen,' Burwen concludes, 'but it gives you some idea of the complexity of the organ Ray Kurzweil is predicting will soon be exceeded by a machine. In my opinion, machine intelligences will not have the generality of their human counterparts, although they may be able to beat humans in certain specialist areas, such as chess or spectrology.'

When we get beyond the structure of the brain and into thinking about what we actually do with our wetware, the prospect of advanced artificial intelligence (at least, that which isn't just a programmer's trick to make a computer *seem* conscious) seems to recede still further.

For one thing, what do we imagine would be the motivation of a conscious computer to stay alive – let alone to have original thoughts? I suggested in chapter two that the obvious, logical response of an intelligent computer to the world's intractable problems would be to unplug itself and commit 'cybercide'.

But even if that response could be programmed out, there is a good case for saying that computers are essentially different from us in that everything we do, from waking up to writing great literature, is prompted by appetites, namely the desire for happiness, physical comfort and sex.

Shakespeare is thought to have written through the night in a desperate bid to keep his creditors off his back. Countless great deeds, from the creation of works of art to bravery in battle to great scientific inventions, have been inspired by love. A book by evolutionary psychologist Geoffrey Miller, *The Mating Mind*, argues that in the case of male humans, most achievements, from art to sparkling conversation, are carried out purely as courtship tools.

While most of us, then, do what we do for money or sex, dilettantes, who work when they don't have to – the nearest motivation it is possible to imagine to that of a robot – are associated primarily with worthless, mediocre endeavours. Perhaps the motivation problem can be overcome by including an artificial vanity program in robotic software. It would be interesting to see such a program in action, however.

But it is creativity, with its essential element of illogicality and unpredictability, that seems the really impossible stumbling block for any machine. 'We long ago became blasé about the wonders wrought by science,' wrote the British political journalist Frank Johnson in an article sceptical of artificial intelligence for *The Spectator* magazine in April 2000. 'We are not blasé about the wonders wrought by humans. That is why most of us are still more awed by the works of Shakespeare or Mozart than by Einstein, much less those of Bill Gates.'

But creativity is funny stuff; on first consideration, it seems to be high flown and have a mindset of its own, above and beyond everyday life. Yet what it actually relies on, according to Professor Michio Kaku, is the heightened level of common sense sometimes called intuition or imagination. Computers have managed to 'write' poetry that looks vaguely like the creation of a human mind, but poetry and abstract art are notoriously easy to pastiche, at least so that the general public might not recognize them as fake.

A novel or a play is a different matter, and George Orwell's novel writing machines, producing pulp for the proletariat, are less likely ever to exist because of the difficulty, for a machine, of sensibly extrapolating from what it knows, to what it doesn't. Creativity requires an internal logic beyond mere machine logic.

'The first fundamental reason robots won't replace humans,' says Professor Kaku, 'is vision. The second is common sense. They don't understand that water is wet, that mothers are older than their daughters, that animals do not like pain. In fact, there are over 100 million lines of common sense that are common to a five-year-old child. What makes us human is our ability to understand common sense, which means that the job opportunities of the future will be those jobs that rely on it. It's impossible to replace artists, comedians or scriptwriters. These are the people who will have jobs, because robots cannot tell a joke.'

Comedy may indeed be artificial intelligence's downfall, not because it is too trivial for computers to be bothered with – but because it is too complex. 'The smartest people are the people who can use humour and appreciate humour,' says educational psychologist Dr Geraldine Schwarz, president of the International Foundation of Learning in Vancouver, British Columbia. 'That, to me, is intelligence at the very highest level, so high that it cannot be defined and explained. It's that different way of looking at things, of seeing things. If you want to know if a person is super-intelligent, just have a look at the humour they use. It's not something measured by tests.'

Creativity or intuition isn't all about comedy or the entertainment industry, either. Marc Demarest, an English literature MA who doubles as a computer guru and chief executive of an IT company in Portland, Oregon, extends creativity to areas such as surgery. 'There is something about the instantaneous decision that a surgeon makes when, during an open heart operation, he nicks an aorta and has to take action, that you will never be able to duplicate with a machine. Never,' says Demarest. 'The human brain is wired as a very sophisticated heuristic or analog computing device. Digital processing is simply not amenable to making multi-variate decisions instantly, because the nature of binary logic is that you have to proceed down branching trees of alternatives.'

So, how intelligent does Demarest believe computer programs will become? As intelligent as we genuinely need them to be, is his answer. 'What's going to happen is that very soon, your software will know you intimately. The first time it is put on, it will ask you thousands of questions. You will spend a long time programming it, even if it doesn't look like you're programming, but it will be learning from you. It might ask, "How do you feel about rain?" and have you move a slider bar. You're wondering, why the hell would it ask me that? But you see, what we're trying to do is not create an artificial being, but to take what today are things which consume significant amounts of human brain power and allow them to be done by cheap digital entities, so that we can do things which are more important. So I don't believe we'll see any of those 1960-ish notions of not going to work. Work will just become more interesting and worthwhile.'

Of course, programmers could try to emulate, along with artificial vanity, the unpredictable *je ne sais quoi* that makes for creativity in computers, but unpredictability without both a moral and a common sense dimension is likely, one would think, to be either downright dangerous – or unintentionally comical. And it would rely on software writers, of all people, to become still more central to everyday life than they already are.

Computer programmers were never really expected to become the default demi-gods they are today. 'Even fifteen years ago, there was no sense that computer technologists were the un-acknowledged legislators of the world as is now the case,' says Marc Demarest. 'It's a little scary because we're not necessarily the most socially responsible breed. In many cases we have bizarre world views. It's amazing the attitudes that are hiding out there, among people who use computers as a substitution mechanism for human relations.'

We have only to look at the ridiculous comedy of errors which computer software currently presents, sceptics may feel, to enjoy a glimpse of the future – and the sound of another nail sinking thankfully into the coffin of artificial intelligence's more extravagant claims.

The software glitch and the computer crash have become the running farce of modern life. They are rapidly turning the very people who are expected within ten years to be the buyers of domestic robots into modern Luddites. Computing pioneer Sir Clive Sinclair commented recently: 'Whenever I'm in a meeting and we're deciding about dates, the people with diaries are always ten times faster than the people with Psions. But the people with Psions get terribly cross if you say so.' But even for those of us at peace with our palmtops, it feels sometimes that if we encounter one more atrociously bad voice recognition robot attempting to book our cinema tickets, we will go mad.

For the most part, software problems have caused only inconvenience or hilarity. Perhaps the definitive case to date, and the template for endless similar fun in the future, occurred in Britain in May 2000, when a phone call from a woman in Aberdeenshire about a mouldy chorizo sausage she had bought in Safeway prompted a national emergency. Alison McKenzie called an environmental health helpline to warn them about the sausage, whereupon a British Telecom automated voicebank system proceeded to pass the alert on to every police station in the UK. For good measure, it also put details of the sausage on to hundreds of thousands of people's pagers.

To believe that the same people whose software can currently cause such a horrendous foul up will in twenty years be writing programs cleverer than the human brain is not easy.

In the rush towards artificial intelligence, it is all too possible to forget that the naked, non-electronically enhanced human brain is bound to remain the thinking organ of choice for the overwhelming majority of the world's population.

Take up of colour TVs, let alone brain implants and robots, is a tortuously slow business compared to the actual invention of cool new stuff; a huge number of people in Europe and North America still have old black-and-white sets; most people on Earth have never even seen a telephone; and the World Wide Web is in reality a thinly stretched thing, far from worldwide, with less than 2 per cent of the world actually on it. The number of people disenfranchised from technological progress, furthermore, is growing rapidly; every day, nearly a quarter of a million people are born, 97 per cent of them into an environment where access to food and water, rather than access to modems, will dominate their entire life.

In the technologically advanced nations, we are concerned about information overload. One of the key theories of Nicholas Negroponte, founder and director of the Massachusetts Institute of Technology Media Laboratory and author of the influential 1995 book *Being Digital*, is that the digital, internet world has brought about 'a change in the distribution of intelligence'.

But the evidence is that it's still an extremely uneven distribution. For those of us who are literate and wired up, information pours into our lives like a flood tide we need to shelter from. The more information which swamps us, the less attention we can give it, and the less, paradoxically, we end up knowing. We have to put on information blinkers, ignoring the potentially fascinating material all around us and focussing relentlessly on the one thing we need to know on a given day. We are rapidly acting out a famous short story by Jorge Luis Borges, 'The Library of Babel', in which a library is discovered containing all the information there has ever been or will ever be. Everyone is thrilled by this, until they realize there is so much data there, that it is impossible for anyone to locate what they want.

Information overload is, however, far from a problem among the billions in the Third World who have no schooling, or the urban poor in the United States and Britain who receive a minimal, watered-down education, less concerned with learning than with behaviour control and the avoidance of offence to minorities by cushioning fundamental issues in cotton-wool syntax. Such young people are being ever more marginalized by what the 1970s futurologist Alvin Toffler called 'the transfer from a brute-force to a brain-force economy'.

What, then, is the future for education? And might the inexorable merging of human and machine have a spin-off in education that could at last benefit those who could use a little information overload?

The 1960s were full of proposals for what were then called teaching machines, which, along with language laboratories, were going to bring about a revolutionary automation of schools. It never really happened, but today, the same is being promised again, this time with computers and the internet.

In Britain, Prime Minister Tony Blair and the Department of Education have put their weight behind proposals to install a computer in every classroom, Blair pointing out that an hour taught by a teacher costs £50, whereas an hour of lessons delivered electronically costs just 75 pence, with that figure dropping every year. An experimental system being developed in Manchester to beam a hologram of a teacher into schoolrooms was well received at the south London school where it was tested in 1999. 'We all found the system very exciting to work with,' maths teacher Catherine Darnton of Graveney School in Tooting told the *Express* newspaper. 'Once the pupils got over the novelty and settled down, the lesson felt surprisingly real.'

The system, invented at the Digital World Centre in Salford Quays, uses a fast internet connection to zip the live image of the remote teacher on to a glass screen in the classroom, which gives the appearance of the teacher standing in the room. He or she has a hidden camera to see the class (or different classes around the country or the world) and can make two-way eye contact with pupils. The technology is being seen as a way of enabling schools in the near future to teach minority subjects, and to share teachers in disciplines like maths, where there is a chronic teacher shortage. It could also enable universities to stage lectures by guest speakers from around the world, without the expense of flying them in.

But will private- and Ivy League-type schools in Britain or anywhere else contemplate offering such a de-personalized, cyber form of education to the children of the elite? Or will computer-based education become a cheap get-out for poor, publicly-educated children only, with mini-riots breaking out in unstaffed schoolrooms, while the well-off continue to be schooled under the eagle eye of caring, flesh-and-blood teachers? Perhaps the question barely needs to be posed. It is perfectly conceivable, however, that education-hungry schools in the Third World might be all too willing to take advantage of such electronic teaching, and it could be a technology to take very seriously indeed for that reason.

If we can be sure of one thing about the future, it is that the human mind is on a journey somewhere. Our spleens may be content to carry on in the old way for ever, or until replacement, but the brain's restless nature demands that it keep changing, keep taking on new challenges.

It may now have engineered itself into a position where it achieves immortality, like the classical gods; or where it has effectively sent itself to the scrapheap because the creatures of its creation have become *its* gods. The brain may, alternatively, conclude, when its intelligent machines ultimately disappoint, that there are more things in heaven and earth than are dreamt of in its philosophy, and retreat to a more humble position.

Whichever way, assuming the continuing existence of a habitable planet, the development of the human mind is sure to be the biggest issue for the immediate future. And whereas in most areas of futurology, there is always some doubt as to whether the present is truly a 'special' time, when it comes to the brain and the nature of intelligence, the evidence is compelling that the next fifty years will dictate the way the next thousand years proceed.

Chapter 6

OHM TRUTHS

Home and Work

Scientific and social progress happen quickly, often following an agenda, or so it seems, set by Hollywood and science fiction. The media's excited babble about new gadgets, new medical techniques and lifestyles follows close on behind. But the public's take-up of almost anything new is painfully slow, even in the West.

This take-up lag applies most of all to anything pertaining to humdrum everyday life. While we are willing to try out new medicines, restaurants, cars and holiday destinations, we are deeply sceptical and resistant to any change in our homes, our work practices or the way we shop.

That is why, despite the unending barrage of style advice in the media portraying modern steel and glass houses and ascetic, minimalist interiors, the great majority of new houses in Britain are currently built to echo hundred-year-old styles. When people move into them, they fill them with furniture and decorations which have not changed much in thirty or fifty years.

At work, most attempts to achieve the 'paperless office' of 1980s dreams have failed dismally, as have the bulk of innovative schemes of the early 1990s to create 'hot-desking' environments, where laptop-centred and mobile-phone-based workers grab whichever desk they like in the morning, shift at will into 'soft' thinking areas, and so on.

Similarly, defying a several years'-long campaign of e-commerce propaganda, we have not taken to it at all. A convenient solution is still being sought to the e-commerce delivery conundrum, the paradox that for goods any bigger than can fit in a mailbox or through a letterbox, the very people most likely to want to shop online are rarely at home to take delivery. More likely to be solved are the still more basic problems of e-commerce – the inability to browse on screen as well as we do in a physical shop, and our unwillingness to buy without trying the goods out.

The browsing problem is already under attack by the development of 'intelligent agents', which know our tastes and pre-select goods for us. Amazon.com already has this facility, and has some success in reproducing the kind of directed serendipity we intuitively rely on when in a real bookstore; the site's opening page often seems to have an uncanny knack of bringing to our attention books we didn't know about, but are interested in. It's not uncanny, in fact, but intelligent agent software.

Still under development are what are called 'avatars' – a word derived from the Sanskrit for the manifestation of a deity in human form. An avatar is a virtual clone of our personal body shape which will soon be able to try on clothes on fashion websites. It may not be long now before *real* twenty-four-hour shopping in a mall as big as the planet is routine. You may be looking for a leather jacket and have an image of it in your head – it's not a motorcycle jacket, it's not a flight jacket but that's the best you can say. Knowing what it knows about you, your intelligent agent will crawl around the world to search out what you want, get your avatar to try it, and even be authorized to buy it for you.

Yet knowing that all these enhancements to e-commerce are round the corner, the most optimistic estimates are that 20–25 per cent of our shopping will be done online by 2015. There is probably a fair case for saying that if we had been buying by computer for hundreds of years, the mail order catalogue (no screen! no leads! no batteries!) would today be regarded as a massive innovation, and the walk-in shop, a retail revolution.

Smart gadgets for the home and office appear to be less popular still. Ian Pearson, British Telecom's resident futurologist, has predicted for the immediate future an epidemic of 'kitchen rage' as consumers fulminate against over-complicated appliances and illiterate, unclear instruction manuals. 'There are problems now, but it is going to get a whole lot worse and there will be real conflict in the home,' Pearson says. This, it might be noted, is *before* the new wave of kitchen gadgets expected before 2005, which will connect our appliances to one another through the internet.

One can only imagine the potential for misery and confusion if dishwashers are soon using the web to summon an engineer and lawn sprinklers to confer with the weather service and decide if they need to come on. An internet refrigerator already built by Electrolux will supposedly keep an eye on what is on its shelves, make a note of the barcodes on food you take out and uni-laterally reorder it by dialling up the supermarket – whether or not you enjoyed the item.

It is not explained how your friends will resist the temptation out of mischief to pass the barcode of that ancient jar of pickled beetroots at the back of your fridge a couple of hundred times across the smart scanner and stand by to giggle as the delivery van arrives laden with more beetroots.

There seems no limit to the eccentricity of appliance designers. A 1998 prototype built by a subsidiary of NCR, the automated teller machine company, makes the internet fridge seem a positively conservative futuristic concept. This project attempted to marry the microwave oven to the ATM to produce the first microwave bank, at which you could pay your bills whilst waiting for your supper to heat up.

And yet laugh as we do at such nonsense, we *do* steadily adopt new technologies if they offer the real promise of a more convenient life, rather than the addition of mere techno-clutter. Take-up lag is deceptive because we don't adopt technologies suddenly, but typically after a series of periods of resistance.

We refuse to buy a cellular phone, then cave in and find it quite useful. A year later, we finally read the instructions, learn how to send text messages, and become slightly more keen. By this time, we have overcome our resistance to buying a PC, because we need it for work and home accounts, but can't be bothered with the internet. A year or so later, we become cautious fans of e-mail, realizing it has revived the art of letter-writing and adds a real new dimension to our social life.

Six months or so after that, we explore the web and find one or two sites at which we are tempted to buy goods. Without realizing it, we have become part of the revolution. No wonder IBM feels confident to predict what at first hearing seems unlikely – that by 2005, at least 30 per cent of new household appliances will interact with humans through body sensors or speech recognition technology.

The same creeping, but nevertheless real, take-up of new concepts happens in the world of work. At the moment, when artificial intelligence gurus like Ray Kurzweil and Hans Moravec predict the end of employment as we know it, it sounds preposterous. 'Entire corporations will exist without any human employees or investors at all,' says Moravec of Carnegie Mellon University's Robotics Institute, writing in *Scientific American*. Our descendants, he believes, 'will cease to work in the sense that we do now. They will probably occupy their days with a variety of social, recreational and artistic pursuits, not unlike today's comfortable retirees or the wealthy leisure classes.'

This trend will become all the more marked if nanotechnology ever enables us to manufacture everything we need from atoms at home, and at minimal cost. Science fiction writer James Halperin, who has studied this emerging domestic technology closely, describes how nanotechnology will work, he believes, within hundreds, rather than thousands, of years.

'Our houses will pretty much configure themselves to our needs. The little nanomachines will reconstruct the floors and the walls to your whim. There's no doubt now; it's pure science. I know people that are attempting to build them now. There are no natural laws which would prevent their existence.

'You will take a piece of raw material, say, wood, and put it into something the size of a toaster. And out will come fettucini alfredo. Or a nice steak. There's almost no limit to what these machines will be able to do, and of course they will be just like computers – they will get cheaper and more powerful. I would think they would be built in your house much the way software is downloaded now.'

A technology which makes manufacturing obsolete would seem to have pretty serious consequences for employment. And yet there is a perfectly good precedent for such seismic shifts in the economy in the mass move away from agriculture, which in a century has seen the proportion of the Western population working in the fields go from an overwhelming majority to a mere handful.

'What are all these people doing?' asks economist Howard Baetjer Jr of Towson University in Maryland. 'They're manicuring ladies' toes at boutiques in New York, they're producing Sega computer toys and producing magazines that appeal to tastes and interests nobody could have imagined some time ago.'

So when futurologists predict that, almost entirely as a result of the rise of electronics, home, work and shopping will effectively merge in the coming century, with all three going on simultaneously in our wired, super-automated (not to forget energy-saving) homes, we ought perhaps to be less sceptical than our instincts incline us to be on reading about microwave ATMs and fridges that go shopping.

Apart from the evidence that we are more amenable to shifting technologies than we like to imagine we are, there is another significant indicator of massive change to the way we will conduct our everyday lives.

All over the Western world, pioneering spirits are already living the life which beckons for the great mass of the population. They are creating automated houses into which they integrate an improved, more productive working regimen as well as an enhanced, locally-based social life, thanks to not having to go through the twentieth-century absurdity of commuting.

This new, wired life is, as we have come to expect, different, but not *that* different, from the visions of a domestic future expressed over the past hundred years. The Victorians, as we saw in chapter one, foresaw automation and even the ordering and delivery of goods through 'tubes'. In Switzerland in the early 1920s, Le Corbusier was reinventing the house as a 'machine for living'. The maverick American architect and thinker R. Buckminster Fuller was inventing aluminum houses and domed cities in the 1920s. His patented 4D Dymaxion House was designed on principles still emerging in ultra-modern concept homes. It revolved around a central 'mast' containing its plumbing and other services and was light enough to be endlessly reconfigured and even mobile if necessary.

Home conveniences like frozen food and the microwave oven were the staple of popular domestic futurology from the *Daily Mail*'s 1928 Ideal Home Exhibition to Walter Cronkite's 1967 CBS TV special, *At Home 2001*. By the late 1960s, the silicon chip microprocessor was additionally bringing the promise of robotic control into view. A 1969 children's book, *Computers at Work*, typifies the spirit of the age by reserving its last, most futuristic, page for an illustration of a machine – not unlike a modern desktop PC – with arrows pointing dramatically to the appliances such an electronic brain would control – a cooker, a heater, a washing machine, a TV and hi-fi, a lawnmower and a set of up-and-over garage doors.

Such dreams were quite swiftly acted upon. In the mid 1980s, in robotics-mad Japan, I was shown round an extraordinary 'house of the future' by its progenitor, a Tokyo University engineering professor, Ken Sakamura. The experimental house, in a fashionable Tokyo suburb, was run by 1,000 microprocessors, installed by 126 computer-related companies.

Everything from the lights to the telephones, baths, the stereo, heating and the rice cooker were tied together in a computer network for easy control. Weather sensors constantly measured the outdoor temperature, humidity, air pressure, wind velocity, wind direction, rain and light. The system would open the windows to pleasant breezes, or close them and turn on the air conditioning. If the hi-fi volume was turned up, a computer could notify another computer, which would close the windows to avoid disturbing the neighbours. The window-shutting computer would simultaneously inform the air conditioner to keep the temperature from rising. If the phone rang, a computer muted the stereo. If you wanted to cook a French meal, a how-to-do-it video appeared in the kitchen, and a computer set the oven to the correct temperature. The lavatory measured your blood pressure, analyzed your urine and recorded the information in your medical records. It also washed your bottom.

The *Tron* house, as it was called, was only a success in publicity terms, however. It was formidably complicated; I managed accidentally to summon the fire brigade by touching what I thought was a light switch; the designers, being typical technologists, had unfortunately forgotten to include a button with which the fire department could be un-called. There was more embarrassment when a study group from the British Department of Trade and Industry came to see the house. A member of the party misunderstood the instructions for the lavatory bowl's integrated bidet feature, and ended up with his grey suit trousers soaked. Perhaps it is no surprise that after a rush of media coverage, Tokyo's intelligent house became dilapidated, and was eventually dismantled.

Several years on, the automatic house of the future is finally gaining ground because of a crucial advance: it has a context in the form of the internet. Smart gadgets in isolation are rarely more than gimmickry; but when they are part of a network – as is the internet fridge, in fact, although that happens to be a particularly ill-thought-out and silly idea – they start to make more sense, and Le Corbusier's 'machine for living' begins to be a reality.

Leading the movement towards intelligent houses, not unnaturally, is Bill Gates, who recently moved with his family to an electronic Xanadu on the shores of Lake Washington, outside Seattle. The Gates' home took seven years to build, was inspired by the space station in the film *2001* and is run by 100 computers in a five-room 'brain centre'.

The house senses Gates's presence, and adjusts lighting, heating, music and even the electronic works of art on the walls according to his mood. As he moves from room to room, the house tracks him, re-routing phone calls and ensuring that the nearest TV is always tuned to the station he is watching. Visitors wear an electronic badge which automatically seals off sections of the house where they are not welcome; more hospitably, the badge can be programmed with the visitor's favourite music, TV channels and artistic tastes. On every subsequent visit, the Gates' guest is thereby assured that the right programmes will be shown on his room TV, and that commodious works of art will appear on the digital flat screens on the walls.

Intelligent houses are not always just a multi-billionaire's toy, either. There is a relatively modest home outside Chicago whose owner can open it by phone from anywhere in the world to let in his sister to feed the cat. (Why she could not simply be left a key is not explained.) The German company Siemens has an experimental installation for the severely handicapped in a two-up, two-down terraced house in Barnsley, England, and several British companies are looking at providing intelligent, hi-tech elements to social housing projects. And inside a deceptively regular-looking new four-bedroom detached house in a suburb of Liverpool, England, retired plumber Roy Stewart and his wife Vera have built one of the most advanced intelligent home installations in the world.

The Stewarts' dream central heating installation, for instance, is extraordinarily smart. It is maintained by a modem link to a control centre thirty miles away in Manchester, where engineers monitoring it and other intelligent houses around the country can detect the smallest malfunction. The heating system is eager to analyse and learn, too. It has, for example, worked out that to get the Stewarts' bathrooms to their preferred 21°C (69.8°F) by morning, it will need to come on at 4.20am if the outside temperature is at a certain level, different times if it is warmer or colder. Being such a precision piece of work, the heating is economical too – an illustration of the principle in intelligent houses that what seems to be purely technological can also be ecological. The Stewarts' first gas bill in their new house worked out at less than £1 a day.

The house has a massively complex lighting system that would do credit to a film studio, it includes thoughtful features such as one Roy borrowed from a hotel he stayed at in Thailand: 'There's nothing worse than waking your partner by turning the bedside light on to go to the loo,' he explains. So he has put in discrete, foot-level lights which automatically switch on as you stumble to the bathroom.

The house is also fitted throughout with a central vacuum system, a rarity even at this rarefied level of convenience. All the blinds and curtains are electrically operated; humidity detectors in the bathrooms sense steam and whisk it away before it has a chance to fog mirrors; a special system Roy devised himself heats up the cold water which gathers overnight in the hot pipes, ensuring there's never a wait for hot water. Even the Stewarts' garden is automated. There's a variety of outdoor lighting options for evenings, plus an auto-irrigation system which senses dry soil and waters the flower beds accordingly. Sprinklers are computer-controlled for minimum water use, and set to turn off if the rain sensors detect any natural precipitation.

Home automation is big in mainland Europe, too. A German CEO has installed a system which enables him to access his weekend home in Portugal by modem as he leaves his office in Stuttgart. By the time he gets to the Algarve each Friday evening, the swimming pool is precisely up to temperature, the air conditioning on, and his bath run.

Siemens has an entire arm devoted to putting elaborate intelligent networks into both offices and homes. Their brochure gives some idea of the attention to detail which 'system integrators' aspire to. If you open a window in a Siemens-adapted house, the window contacts can send a message to the nearest radiator to shut down, no energy is wasted heating the outside world. When you leave home, the system can be made at the turn of one key to power down all the lights, activate the security and give a warning bleep if a window anywhere in the house is open. The same installation can be made to ensure that the dishwasher only starts its programmed cycle when the temperature of the water in its solar-power collector is high, or the electricity tariff is at a night-time low.

While it is clear that this kind of home automation still falls into the rich man's toy category – at a house in the Rhein/Main area of Germany, the banker owner requested chandeliers which could be electrically lowered for cleaning – there is today a crucial new vigour behind the movement to automate domestic tasks. What is novel is energy conservation; it is no longer so easy to dismiss as flippant someone with retractable chandeliers, if they also have their bathwater automatically recycled as irrigation for the lawn, or have a heat exchanger in the loft suck out the residual warmth from stale air and feed it back into the hot water tank.

Greenness is both the value-added ingredient to home intelligence and its new intellectual strength. The intelligent home has been recruited as part of the struggle to save the planet, and from that new moral high ground, has gone on to become an agent for social good. Energy conservation is an ethical duty for the rich, but also saves vital cash for the poor. And if a house can be wired to motorize chandeliers or provide hi-fi music from hidden speakers in every room, the same wires can power aids for the disabled or the means for the family to keep an eye via a video link with a bedridden granny upstairs – or to connect a home office in any spot, however remote, to the outside commercial world.

Osgathorpe, a village of 470 people, which nestles in a pleasing fold of countryside in Leicestershire in the English midlands, is home to an archetypal new-wave business of the type which could regenerate villages and rural communities the world over.

Nancy Slessenger is to all outward appearances a typical professional woman of the type who has given up work to raise a family and live a classically British country life amidst the farm smells, the wisps of coal smoke, the sounds of distant crows and the horse riders clip-clopping by the Royal Oak pub.

Nancy lives in a handsome Victorian village house with a five-year-old daughter and a successful husband who commutes daily to the nearby town of Loughborough. There is little even on entering the Slessenger household to betray the reality that Nancy is pursuing a management training and development consultancy career full time from home – yet she is doing so entirely online as part of the new 'networked economy'.

Nancy's company, Vine House Essential, is an early prototype for what are becoming known as 'diaspora companies', in which colleagues live precisely where they want, while remaining very much part of a team, with shared team goals. It consists of six women and four men in their thirties scattered in villages around Britain. Nancy's personal assistant is a hundred miles away in Suffolk, with other members of the team in Nottinghamshire, Warwickshire and Northamptonshire. They have no office, and meet no more than a couple of times a year.

Nancy Slessenger started Vine House – the company is named after her home – when she became convinced that the majority of time spent in offices, as well as travelling to them, is wasted. 'There are plenty of people in Osgathorpe who are probably unaware that I work at all,' she explains. 'One of the many beauties of working from home is that you can slot in things like baking or popping out to the organic farm to buy food.'

One of Nancy's partners tells similarly how she has got to know her Northamptonshire village. 'I used to work in the City of London, but now I've got a proper life,' she says. 'I know people in the village, the postmaster and so on – and at the same time, I've quadrupled my income since moving out.'

One of the classically bizarre touches which no diaspora company can seemingly be without, is that the engineer who for the first two years of the company's life maintained its computers was some way distant – in Sydney, Australia. The twelve-hour time difference enabled him to trawl through the partners' systems and sort out any glitches while they were asleep. With such technological near magic becoming routine in rural spots, the concept of the global village truly begins to come to life – as does the truth of Arthur C. Clarke's view that 'any sufficiently advanced technology is indistinguishable from magic'.

David Stewart, a Scottish entrepreneur, has gone perhaps one better still in his flight from the traditional office. Stewart's companies are all physically located in Kiev, Ukraine. He is a joint owner there of an FM rock music station, an advertising agency, a billboard company, a media-buying network with offices located throughout the Caucasus and central Asia, a media sales house, a Russian-language internet portal, a web design company, a market research organization, and the Ukraine's major ice hockey team. Such an empire takes a lot of running, yet Stewart only goes to his Kiev office once a month.

Sometimes he runs his Perekhid Media Group online from his laptop in an Edwardian house in Dunblane, central Scotland, but even there does not have anything resembling a permanent workstation. If the mood takes him (and he takes the laptop) he might work from his sixteenth-century house near Fort William in the Highlands, from his wife's lake house in northern Wisconsin, USA, or from a hunting lodge in South Africa.

Technology means that David's whereabouts are almost immaterial to his staff, as their e-mails – as many as a hundred a day, plus correspondence with his bank on the Isle of Man – reach him instantaneously anywhere. He may not reply immediately, if, for example, he is walking the dogs in Scotland or out in the bush in Africa, but to all practical purposes, he is able to be at work and at leisure at the same time.

Such businesses as David Stewart's and Nancy Slessenger's are by no means anomalous, either. Almost every village in the Western world seems to contain a wired worker or two. Most significantly, though, the talk in small towns and villages is all of modems, e-mail, ADSL lines and websites. The revolution is spreading slowly to the Third World, too. One rural Scottish company which sells Scottish products on the internet to the United States employs a marketing specialist who lives in a remote village in China.

The development of remote working from electronic cottages has been of particular satisfaction to information age guru Marc Demarest, in Portland, Oregon, as he has been forecasting for several years that such new working practices will lead to permanent geophysical and social changes across the world.

Demarest is a thirty-nine-year-old polymath with an IQ hitting the end scale (it was last measured at 182) and a degree in political science and nineteenth-century English literature. He became interested in computers through his desire to compare their textual subtleties with those of original manuscripts, was later a computer manual writer, then an engineer. He now concentrates with his company, DecisionPoint Applications, on helping businesses build data warehouses to manage their affairs based on the historical record of the business.

Demarest regards the industrial age, when your home was dictated by the location of your work, as an anomalous social development. 'Rip apart the coincidence of home and work and everything will change,' he says. He is confident that home working will lead to the density of population smoothing out, to environmental advantages through a drastic decrease in pollution, and to a rebirth of both rural communities and of the stability of the family unit, the latter as a result of parents being at home a great deal more for their children. He also predicts the internet will be responsible for a return to agrarian pursuits, especially growing home produce.

Demarest goes on to imagine the development of communities based less on the need to be close to a particular factory or office, but more on shared interests. 'This thing will mean we will start seeing people living in a community of politically like-minded people, in golf communities, retirement communities, towns of people who like to read.

'One of the driving binary oppositions of the twentieth century was work versus home,' he says. 'Yet that didn't exist until the early decades of the 1800s. It didn't matter whether you were a shopkeeper or a farmer or a textile worker – you did what you did in your house. There are fascinating pictures from the first illustrated books in the UK in the nineteenth century of kitchens which were also clearly industrial production spaces. But as commerce scaled up, we took these disparate workers and put them all in one physical space so that we could scrutinize and routinize them, and the factory was born.'

Now, however, along with other social historians, Demarest sees the core of commercial companies getting smaller, and enterprises being surrounded and serviced by a constellation of freelance, semi-employed speciality workers. 'When we finally break away from the physicality of work – "I must come to this office and sit in this chair in order to, a) use technology, b) talk on the telephone, c) have a video conference" – then we bring back the possibility of re-uniting work and home,' he explains.

'It strikes me that the core of a firm – ten, twenty or fifty people – can probably never be solely associated with one another by electronics. For that core, being able to hug another person, or punch them in the arm, or see the look on their face, is crucially important to the soul of the firm. But in my own life, when I was until recently a senior executive in a fairly large computer company, I could still spend about 75 per cent of my time not physically co-located with the people that I worked with. The technology was already sufficiently robust that I could do most kinds of work – see them, talk to them, co-ordinate with them – using a variety of mechanisms, all of which were in my house, twenty-five miles from my plant. I also spent about 30 per cent of my time with customers, who were all over the world, which meant I had to work from hotel rooms and airports.'

With internet technology virtually perfected, Demarest continues, the social issues are the remaining challenge – and the concept of the fully distributed firm, he is convinced, will become socially acceptable in the next generation. 'My former chief executive officer was in his late fifties, and if I talked to him about this notion of a firm which exists only electronically, he became violently angry, even though he is a very serious user of the technology. He believed such a model radically understates the importance of physical co-location, of viscerality. He is a person of the body. I am a generation behind him, and I see his point. But my son, who is eleven, doesn't care. There is nothing about his experience of the electronic universe which is less visceral or less complete than his experience of the physical universe. His generation is probably the first for which viscerality will no longer be an issue.'

This non-physical level of connection may seem to some to be a psychologically troubling prospect. But Demarest is starting to believe that the wired world could increase the stability of families. 'My son and his grandfather – my father – live on opposite sides of the United States, but communicate regularly with one another in this electronic multi-media environment, which is as fully gratifying to both of them as a telephone call. The web is being talked about as a new kind of habitation, and I have a group of people, some of whom I haven't seen face to face for ten years, but I correspond with on a daily basis. So I think it highly likely that we will see a resurgence of more of these older, communal forms of relationship in which the nuclear family is still a viable element, but in which children are actually raised by a community of people in which adult members don't see themselves as someone's husband or wife or father or mother, but as members of a community that have a shared set of values.

'My primary sense of social identity,' he argues, 'is from the people I work with. But I believe my children's generation will have a different set of criteria for choosing where they live, and it will be thematically orientated. I can see very clearly communities occupying dying village infrastructure in the US or the UK, and these communities being united in their passionate interest in golf, music, or a certain fringy religious belief.

'So when in the future, I am living in a commune in Oregon with fifty other people who are passionately devoted to obscure Victorian novelists, *that's* where I will derive my primary sense of identity. I am not going to feel any need to punch my employer in the arm, or hug him. At any time, I am intensely loyal to the company I work for,' Demarest concludes. 'I routinely make decisions that damage me as a human being in the interests of my firm's interest. But my child will not do that. He will consistently give loyalty in the direction of himself and his community rather than in the direction of employment, which will become much more of a contract orientated relationship.'

Embroidery communes in Oregon and stamp-collecting collectives in Cornwall aside, what will our towns and cities look like in years to come? One of the most curious things about futurology is that in a hundred years, visions of the future broadly like that of the 1939 Futurama pavilion at the New York World's Fair (see chapter two) have barely changed.

The same fixation with height and multi-level transport and living applies today as when the first ten-storey 'skyscraper' appeared in Chicago in 1885. For the rather obvious reason that helicopters in their thousands would bump into one another if they were allowed to buzz like flies from one skyscraper to another, city transport has remained a street level or beneath street level activity. Because of the obsession with aerial city transport, it was thought earlier this century that tall buildings would all have their main entrance on the roof, but this has manifestly not been the case, and it can be safely assumed that everyone other than CEOs and rock stars will continue to be strangers to the rooftop helipad for a very long time to come.

Perhaps there is something phallic about tall buildings that makes architects continue to construct them higher and higher, but beyond accepting that city centres will become ever spikier and more dramatic, there is no reason to imagine that skylines will look substantially different a hundred or more years from now.

What will be more interesting to watch will be the suburbs, which cover the majority of urban land and house most people. Would we feel alienated and disoriented if we could leap ahead to the suburbia of 2100? Or will the streets look superficially similar?

In the greater part of the United States, South America, Japan, Australasia or Scandinavia, the likelihood is that people's sur- roundings will look remarkably similar even two hundred years ahead. Houses are newer and better built, and, earthquakes notwithstanding, should survive in reasonable shape. The innate design conservatism of the vast majority of people should ensure that they are not demolished and rebuilt along Buckminster Fuller or Le Corbusier lines merely for the sake of progress.

Europe, however, begs a different question. The building stock of the UK, for instance, dates largely from the two vast construction booms of the late nineteenth century and the 1930s. France, Italy, the Low Countries, Eastern Europe, Spain and Germany all have huge stocks of housing and commercial buildings of seventy to 150 years' vintage, albeit brim-full of the latest high-tech electronics. Is it all likely to fall down? Or do modern building techniques guarantee old buildings a virtually endless life? And if our Victorian and 1930s houses do become uninhabitable, what will replace them – especially when lack of space and high land values dictate that a suburban terrace of housing in Europe, or the kind of 1930s row housing seen in older parts of the United States, is probably destined to remain a tightly bunched series of linked properties?

'Nothing is forever,' explains Professor Rodney Howes, head of the Department of the Built Environment, at London's South Bank University. 'It is generally reckoned that the normal life cycle of a dwelling is somewhere between sixty and a hundred years. Now, there are silicones and various hard, transparent coatings for existing materials to make them a lot more durable and less prone to deterioration. It's like embalming people, and within the field of conservation that will be a growing area – if that is indeed what society wants. So if you spend the money on conserving the structure, then buildings can last for hundreds of years.

'But the point is, these buildings aren't energy efficient, and if you don't have the energy to heat homes, then there is another driver which forces you to go down the replacement road. So I would say in the main, a lot of the Victorian housing in British cities will go. Putting new roofs and windows on them and cleaning up the outside brickwork can extend the life of them by another seventy or eighty years maybe. The only way of making them energy efficient is to lag them *inside* with heavy insulation, but that reduces the floor area.

'It will be important to retain heritage and preserve the best of what we have, but at the same time we have got to look to changing needs and new and innovative material which really make our homes affordable and far cheaper and easier to use and give a far better functional performance. I can see the use of new and innovative materials that can actually be recycled from what we throw away at the moment. People involved in afford-able housing in China, for instance, are talking about waste wool, which is stuff that nobody uses, but if you compress it and add acrylic adhesive, you get a board material which is very light but as strong as carbon fibre.'

The question of what new suburban housing might look like in 2100 is, of course, as unanswerable as predicting tomorrow's fashions. Given the existing space constraint of the suburban street, however, it is fascinating to try to envisage what a new terrace of twenty-second-century houses would be like. Design conservatism, again, may mean that the builders of that time will still be mimicking the 200-or-more-years-old houses they replace.

It is sobering indeed to think that a hundred years from now, even if they have been rebuilt, the residential streets of London, Manchester or Philadelphia might well look reasonably familiar to a visitor from the nineteenth century. There are very likely more ways than we imagine that the future will be surprisingly similar to the past.

Chapter 7

TIME OUT

Leisure

A modern parable tells of a British missionary in a Third World country coming across a young man fast asleep under a palm tree in the middle of the day. 'What are you doing, man?' he shouts at the prone figure, 'get up, get working. A fit chap like you should be earning money and making a career for himself, not lounging around.'

'Why?' asks the man as he tips his hat up to talk to the Englishman. Apoplectic now, the missionary launches into a sermon on the work ethic, and how it benefits people by helping them have a home, buy things for their family, and so on.

'I'm still not quite sure I understand your point,' the man says. 'I have a home and I have everything I need for my family by doing the minimal amount of work I already do. Why would I want to earn more money?'

'Well for a start,' splutters the Englishman, 'it would give you the chance to get in a bit of leisure time.'

An observation by one of the leading figures in the modern leisure industry tells a similar story in a different way. Andre Jordan is a sixty-seven-year-old Polish-born former journalist based in Rio de Janeiro, from where he runs several property development companies. He is the founder of a clutch of leisure resorts in Portugal and Spain, including Quinta do Lago on the Algarve, and nearby Vilamoura, the largest resort community in Europe.

'What we have done in these resorts is to provide a leisure

version of traditional life,' Jordan says, explaining that the wealthy families who have homes there are usually spread all over the globe, and typically gather in their house on the Algarve for official family holidays and for Christmas.

'However,' Jordan continues, 'I'm not at all sure that the ease and pervasiveness of technology has made people's lives necessarily better. When I'm at Quinta do Lago, I go out for long early morning walks, and I see these young, hotshot bankers or brokers whatever they are, jogging and talking on their mobile phones at 6 am. Now to me, this seems self defeating. The availability of communication seems to have invaded people's lives to an unacceptable level.

'Is the technology serving them, or are they serving the technology? When I was a young man, when you were at your holiday home you were incommunicado; when you were at the beach you were at the beach and nobody could reach you, and that was best in all ways for you and your business, that you relax. But now you are at the beach and your phone rings. It is so common to see people by the sea or on the golf course talking on their phones that we don't even remark on it. And faxes and e-mails eat into all-important leisure the same way.'

Leisure was not always supposed to be this implausible two-way stretch between work and relaxation. If there was one question on which the futurologists of the past hundred years have concurred, it is that by 2000, we would barely be working at all, but wallowing in free time, and most probably forced to devise all kinds of inventive new ways of filling those expanses of time.

The Victorians expected us to extend our search for something to occupy our minds into the air and under the sea. Air tennis, air racing and fishing from balloons were confidently forecast, as were undersea croquet, fishing from submarines and fish racing, in which the jockeys would apparently sit on the backs of thoroughbred cod. (They also forecast that fox hunting would become a motorized activity, with springs under a car's chassis powering the vehicle over the jumps.)

Needless to say, the Victorians had it semi-right – computer games are surely the exemplar of a pastime invented to pass the time. But something quite unforeseen has happened; in 2000, we work much longer hours *and* spend vastly more time on recreation. This might suggest that the day has been lengthened in some way, and to an extent, it has; electric lighting led to the extension of daytime activity into the nights, and that in turn to the twenty-four-hour society we now enjoy, sleeping far less than we did a hundred years ago.

The modern increase in the time we spend working began as a response to job insecurity in the 1980s after a steady 200-year decline in hours. It is very likely that it will ultimately go into reverse, and the early futurologists will be seen to have, as so often happens, got the trend right but the timing wrong.

But in the meantime, we are left with the puzzle of what has given way to make time for us, at the same time as working harder, to have become the weekend mini-breaking, mountain biking, internet-surfing, gym-haunting, channel-hopping creatures we now are.

The British Office for National Statistics reported in 1999 that for the first time, households were spending more on leisure than on housing, food or transport. It is tempting to say all this extra spare time is explained by the fact that we read and study less, but in reality, we seem to do both far more, with book and newspaper buying booming. Letter writing, at least in the form of e-mail, has also grown hugely. Dr Michael Argyle, the veteran Oxford University social scientist, suggests tartly: 'I tell you what people are not doing. They're not seeing their friends. They're seeing imaginary friends instead, in TV soap operas, and it's rather disturbing.'

Dr Argyle is probably right to an extent. But what has surely diminished more would seem to be the proportion of the day we spend on household chores. Running a home – from laying fires to scrubbing doorsteps to sweeping to cooking – has compacted down from a twelve-hour-a-day occupation to something which need not take up more than an hour or two.

The shiny-eyed, futuristic-minded proponents of electricity promised from the turn of the twentieth century onwards that it would reduce drudgery, and it did precisely that. The main reason that the prospect of household robots has lost its appeal – indeed, why the idea of having a domestic robot strikes most of us today as an *increase* in drudgery, with yet another instruction book to read and another machine to have a software seizure – is that there's not that much to do to maintain a modern home.

It is really kept running by a platoon of little, *un*intelligent robots, from the gas central heating controller to the oven timer to the answering machine, but we don't tend to think of them as bots, just electrical conveniences. Add to the time and effort saved by these, the enormous liberation afforded by convenience foods, supermarket shopping and deep freezers, and it is easy to see how leisure has come to take up such a large slice of our lives without having made any inroads at all into our expanded working hours.

Problem-making animals that we humans are, we have naturally managed to turn leisure into the opposite of itself – a pressure. Just as those who have plenty of money, and ought therefore to have no worries, say that the bother of tending and spending it actually becomes stressful, a surfeit of leisure manufactures as a by-product a special kind of angst.

Perhaps it is the insidious nature of the late twentieth-century work ethic, perhaps the fault of universal education making us ever more hungry for mental stimulation, but seeking the *right* leisure activity, which is constructive, educational and makes us mentally or physically fitter, has become a preoccupation.

We watch garbage TV, but most of us feel vaguely guilty for doing so, worrying that 'chewing gum for the eyes' will make us less able to keep up during the next workout for the brain. We lounge around swimming pools on holiday, inwardly fretting that we ought to be feasting on works of art in the nearest museum. We eat supermarket pre-prepared dinners, while feeling increasingly threatened by the growing pile of fashionable and lavishly illustrated – but unopened – cookbooks on our shelves, and the piles of recipes carefully torn out of Sunday newspapers a week or ten years ago, but yet to be tried out.

Hearing about studies such as one in Sweden in 1982, which concluded after surveying 12,000 people that educational activities make us live longer by stimulating the immune system, we have even started to compress leisure down to make time for more leisure.

In 1999, I was introduced to the concept of the yuppie trek, whereby a group of fifty professionals left London on a Thursday evening, flew to Morocco, climbed one of Africa's tallest mountains, and were back at our desks forty-eight hours later, mentally stimulated and having achieved a real measure of confidence and self-knowledge, but at the same time, aching all over and needing a proper holiday.

The High Atlas mountains are infested with such fast-track trekkers, doing in a couple of days what the Victorian gentleman traveller would have spent weeks over. The Inca Trail in Peru is said to be even busier, with those trying to find themselves, but *quickly*, traipsing in long lines not dissimilar to an urban freeway in the rush hour. And anyone wanting to row their own raft down the Grand Canyon rather than buy a ticket for a commercial trip has to face a twelve-year wait for a permit, such is the growing number of people wanting to get the most intense recreational experience in the fastest time.

The perfect metaphor for the modern desire to cram leisure into available time will, for Martin Hayward, director of the Henley Centre for Forecasting in London, be the ascendancy in the near future of the nine-hole golf course. 'Eighteen holes will be seen to take too much time,' he told London's *Sunday Times*. 'The one-day cricket match will replace the five-day test. Sunday morning slouching will be replaced by a more intense approach to leisure, whether it is weekend breaks or high-activity sports.'

Hayward's centre has coined the phrase 'leisure canapés' to describe what it believes will be the new trend in 'surfing' through mini-versions of several different pursuits. It envisages people going to leisure multiplexes where they will exercise, check out the internet, see a movie and have dinner under the same roof. (At the same time, it predicts a boom in 'chilling' – the adoption of some kind of deliberately induced vegetative state as a way of achieving the relaxation we will have lost sight of in all this frantic pursuit of leisure.)

Yet despite such perceptive-seeming forecasts, trying to predict leisure trends is almost – if not quite – as fatuous as second-guessing fashion more than a season or so ahead; the element of whim in leisure is too strong for even straight-line extrapolation of current trends, which normally gives a clue as to how the future will work out, to be very useful.

Who, for example, would have imagined that the mountain bike would grow in a few years from being the passion of a few rough-track hill-cycling enthusiasts in northern California to becoming the bicycle of choice for people living in cities the world over, with hardly so much as a gentle slope in sight, let alone a mountain? There was no compelling reason for urban cyclists to adopt the mountain bike other than that they looked cool on them. And so it is with most leisure trends, driven as they are by incalculable fashion factors.

There are notable exceptions to this rule. It is a near certainty, for example, that golf, and not just the truncated nine-hole version, will be an enormous boom industry for the foreseeable future. The game is reasonably healthy and social, but most importantly, people can play for practically as long as they live. Golfers in their nineties are not a rarity today, even before the anticipated increase in average longevity into the hundreds.

Similarly, we can be almost certain that pet ownership will become a huge preoccupation, as people have children later, and also strive in an increasingly technological world to have some kind of remaining contact with nature. Of course, twenty-first- and twenty-second-century domestic animals may not be just for Christmas, nor merely for their own lifespan; we could be cloning pets, so the day Spot the dog dies and the children are devastated, an identical Spot puppy will be conceived from the old boy's genetic material, and delivered within a few weeks, with the identical markings, temperament and fetching little ways.

Inspired by the cloning of Dolly the sheep, a handful of companies and academics are working on pet cloning. PerPET-uate, a Connecticut enterprise, is storing DNA for pet owners in the expectation of imminent cloning technology. Scientists at Yamaguchi University in Japan are said to be close to cloning a cat. And a San Francisco company called BARC – Bio Arts and Research Corporation – is working on a project called Missyplicity – the prototype cloning of Missy, a thirteen-year-old spayed mongrel belonging to a mystery Silicon Valley couple.

BARC's co-founder, Lou Hawthorne, who also has a stake in a new commercial pet gene bank called Genetic Savings and Clone, believes the cost of cloning a pet could soon be $20,000 or less. 'We will probably have as many orders as we can handle,' he told the London *Sunday Telegraph*. 'Never under-estimate the love some people have for their pets. Shortly after Missyplicity was announced, I had a call from this guy. He said, "I don't have millions, but I could manage a small amount." He was prepared to spend $200,000. And this wasn't a particularly wealthy man.'

There will be no real limits to the areas we exploit for leisure. We will look both outwards, to the world and universe around us, and inwards, to find more to do with our minds.

In accordance with the Victorian futurologists' obsession with mankind extending his domain underwater, there is probably a very good chance that we will start to use the oceans far more as a leisure resource. It is a growing complaint of adventurers that there is practically nowhere left on Earth that can be called a wilderness, yet 99 per cent of the planet's space, in the form of oceans, remains unexplored. Along with increased farming of fish, we can very likely expect an increase in scuba diving and even submarine cruises – unless rampant claustrophobia nips that travel prospect in the bud.

Another almost guaranteed leisure pursuit for the twenty-first century with unlimited growth potential is that ultimate time-filler, the computer game. When a young engineer called Nolan Bushnell brought out the first commercial video game, *Computer Space*, in 1971 – it involved a battle between space ships and flying saucers – it sold only 2,000 copies. But when, two years later, Bushnell went on to invent a table tennis-based game, *Pong*, it captivated the public. Its first installation was in Andy Capp's Tavern in Sunnyvale, California, where the machine broke down after a few hours because it was jammed with the coins of so many customers who wanted to experience what we now call interactivity – the strange new sensation of pitting their skill and co-ordination against a computer. Bushnell founded a company to market *Pong*, and called it Atari; by 1974, 100,000 copies of his game were selling, and in 1982, Bushnell sold Atari to Warner, and pocketed $15m.

From there to modern games like *Myst*, *Tomb Raider* and *Crash Bandicoot* might seem to be a fairly straight evolutionary line. Yet computer game experts say the genre has only just started out, and will soon be an entertainment which truly rivals TV and cinema, both in terms of engaging the audience's emotions and becoming, or so they say, high art, authored less by computer nerds and more by real artists, or computer game directors, as they might well become known.

For those who are rather proud of never having played a videogame since *Pong*, the extent to which computer entertainment has already developed a cultural life of its own may come as a surprise. Take this extract from a review of the game *Riven*, by Tamara I. Hladik on the American SciFi Channel's website, *www.scifi.com*.

'Players assume the role of an unwitting traveller who has stumbled into a mysterious world that is the portal to others,' Hladik writes. 'She/he agrees to help Atrus, the kindly, wise magician-inventor who is the creator of much of the wonders in these worlds. Players are sent by Atrus to *Riven* in the hope that they might find his wife, Catherine, imprisoned by the unscrupulous despot, Gehn. Along the way there are a myriad of switches to flip, cranks to turn and hieroglyphs to decipher ... Entire populations of technologically-simple people are oppressed by Atrus' nemesis, Gehn. Evidence of Gehn's cruelty is in his machines and inventions, and these clues help to foment a rescuer mentality in players. *Riven* also has critical junctions, where certain actions take place that drastically affect the resolution of the game ... *Riven*, like *Myst*, is a staggering success, soldering gadgetry and mysticism together in an indefatigable argument for distraction.'

The idea that computer games are becoming art may still sound an extravagant claim, even with the evidence of such a cerebral level of criticism. And yet it is undeniable that film took nearly half a century to develop from a novelty to an art form. As in the movies, where there could be no substantial story telling until the invention of the soundtrack, or cartoons, where the seaside What-The-Butler-Saw Machine ceased being a mere gimmick only after it morphed into *Bambi*, it will be technological improvement which lifts the computer game into something which could conceivably be the centrepiece of an adult night out.

Another development which seems surprising, but shouldn't be so on reflection, is the emergence of computer game playing as a professional, spectator sport. Networking of games on the internet, by which other people and their various mental strategies and approaches become a part of the playing experience, has for a few years been widening the computer game out from what was described by the British writer Tony Parsons as 'young men sitting in their stained, darkened bedrooms', to an activity more akin to chess, in that it requires vestigial elements of social skill.

Now, however, pro-gaming for money in organized tournaments is becoming commonplace, with prizes up to $100,000 regularly on offer from the US-based Cyberathletes' Professional League. TV coverage of these competitions, complete with replays and expert analysis, is also in development. One such proposed show by the BBC in Britain, which will feature celebrities playing videogames against one another, is provisionally called *Bleeding Thumbs*.

In what may be the start of a coming trend, one of Britain's leading pro-gamers, Sujoy Roy, gave up a 'boring' job as an investment banker with J.P. Morgan to concentrate on his game. Roy and another of the country's finest, Amir Haleem, who won the $10,000 first prize at a Dallas computer game competition in 1999, now run a 'gamer lifestyle' website, *xsreality.com* ('Where gaming meets reality') from Stockholm, Sweden.

A sad degradation of the Olympian ideal of sport? Far from it, according to Steven Poole, author of a British book, *Trigger Happy: The Inner Life of Videogames*. Poole points out that computer games don't segregate men from women or the able bodied from the disabled. One of the top players at one UK tournament, he points out, was disabled. 'A game of multiplayer *Quake 3* or *Unreal Tournament* is actually a revolutionary democratization of the nature of sport,' says Poole. 'Laurels are no longer determined by the tyranny of genes.'

Not that 'real' sport, involving the shameless flaunting of athletic genes, shows much sign of disappearing in the next century. Although the prospect was raised in an earlier chapter of traditional sport soon degenerating into a competition between different genetic engineering companies' skill at enhancing natural human abilities, there is probably still plenty enough scope left in the pitting of human body against body for such early science fiction spectacles as robot-on-robot gladiatorial contests to be a non-starter.

Sports fixtures, like live theatre and concerts, will probably continue to look much as they do now for the imaginable future, even if it is just possible that we will be watching teams composed of modern prodigies alongside clones of long-deceased but legendary athletes.

The lure of *seeing* something happen live will remain because it offers the chance to feel that we are not only witnessing an event, albeit in the case of theatre an *ersatz* one, but meaningfully participating in it by being there, cheering and booing and sharing emotion with tens of thousands of other popcorn-eating people.

Both sports and theatre are arguably best seen on TV, yet the live performance looks set somehow to remain the more prized experience. In the case of live sport, there's the chance of seeing something spontaneous and unscripted, but we don't seem to mind even when we know the ending, as we so often do in drama.

This odd human capacity to submit emotionally to what is really a scripted illusion reaches its most curious in cinemas. If the drama we wish to see has been preserved on film and the actors long since gone home, there would seem to be precious little reason for wanting to watch it through the heads of rows full of coughing, whispering, candy-wrapper-rustling people.

It is not that cinemas, even the most advanced, offer a technical experience unattainable elsewhere. Television, then video, and now Digital Versatile Disc (DVD) and home cinema ought to have made cinemas a wholly redundant anachronism – just as radio *should* long ago have decimated newspapers. DVD and some of the elaborate home cinema equipment available even today offer pictures and sound reproduction which have the somewhat paradoxical quality of appearing more real than reality. And yet we go to cinemas in increasing numbers, paying often extortionate ticket prices to do so.

To understand why, it is only necessary to see a London cinema audience whooping and cheering in their hundreds at the end of a screening of *Four Weddings and a Funeral* – or to see *Schindler's List* at a cinema in Leipzig, Germany, where the audience all bought buckets of popcorn before the film and left them totally untouched for its entire duration. Other people may have been hell for Jean-Paul Sartre, but for most of us, their presence seems to be an integral part of experiencing emotional scenes in drama, as well as sport.

Of course, even if a live show remains most people's big treat, the advance of new and enhanced ways of seeing sports and drama at home is bound to be inexorable. People may shy away from live sports events if the disturbing trend towards violence amongst spectators in almost all sports all over the world continues, but at home, you are safe and available to be entertained unendingly – and TV will provide an ever more amazing show.

To the generations that grew up never knowing what it was like not to have a remote control, interactivity will seem as natural as grammar. Personal point-of-view digital TV, which is already in operation, will improve to the point where the viewer can focus on any part of the action he chooses, from the argy-bargy in the players' tunnel at a soccer game to the cockpit of a Formula 1 racing car.

Then there's the ever-threatening, but to date never-quite-happening, question of virtual reality and how it might boost sports coverage. The very term 'coverage' might come to sound quaint in the age, perhaps fifty years from now, of total TV immersion, or what might be called Vicarious Reality.

A neural input unit something like a swimmer's cap, it is envisaged, will send signals into the viewer's brain. He will watch a soccer free kick from the kicker's point of view and feel in his leg the kicker's sensation of booting the ball; equally, he might feel the exhaustion of a long-distance runner, transmitted directly from the runner's brain and muscles; or experience as if it were his own, the thrust and pain of an upper cut from the heavyweight boxing champ of the day.

Well, maybe. In a world where bungee jumping, yuppie hobo-ing, phone-card collecting and toe wrestling are acknowledged leisure pursuits, whereas, in Britain, an apparently obvious winner such as CB radio was stillborn through lack of public interest, it is manifestly suicidal as a futurologist to predict any leisure trend beyond the more-or-less obvious.

Nevertheless, in the possibles bracket for our leisured future, come such trends as steadily increasing gambling. National lotteries with ever-larger prizes have made financial thrill-seeking an everyday activity, and there is no reason to imagine that prizes will get progressively bigger. The internet has brought gambling to parts of the world, especially the Islamic countries, where it would not otherwise exist. Some airlines have begun installing small-stakes interactive gambling systems as entertainment.

The catalyst for a truly enormous increase in gambling globally could be the internet taking off in China, where there is a vast pent-up gambling fever. If and when China abandons the vestiges of communism, it very well might become a global gambling centre.

Traditional forms of culture are most likely to continue their upwards trend for the conceivable future, very possibly garnering increasing snob value for the middle classes, as they become anxious to distance themselves from the newer, more technological and artificial intelligence-based forms of entertainment.

The young intelligentsia may gravitate for novelty's sake towards 'world music' from local groups in places like Outer Mongolia, Eastern Europe and Africa. But music for the mature middle class will not necessarily become fossilized. Just as percussion and electronic music established a presence in orchestral work in the twentieth century, some new instrument will doubtlessly emerge in the twenty-first. Opera may well move with the times too. 'I can envisage a wave of fine new operas being written in the next century,' says Sir Peter Hall, a former director of the British Royal Shakespeare Company, the National Theatre and Glyndebourne. 'But I can't say when or why it will happen, just as no one can really say what produced the revolutionary surge of Italian opera in the nineteenth century, or what produced Shakespeare. It's just the way the cards fall.'

Another just-possible leisure trend will be for restaurants which serve a social purpose while keeping actual food consumption to a minimum. We have already seen an upsurge worldwide in the kind of minimalist, Japanese noodle-slurping restaurants which were always vaguely hinted at in mid twentieth-century science fiction as a trend for the year 2000.

The next Big Thing may have been heralded by a recent experiment in 'conceptual dining', in Tel Aviv, Israel, called the Café Make Believe, and serving food as near to content-free as possible. Eel mouse and pomegranate salad were among the delights offered by Swiss chef Philippe Kaufman. Not to be outdone, another restaurant soon opened on the city's fashionable Sheinken Street which served no food or drink at all, and still managed to attract crowds of 'virtual' diners. The second restaurant later turned out to be an art students' final year project, but the possibility for an increasing number of extremely unorthodox eating places to appear in the future should not be dismissed.

Similarly, while virtual reality holidays may or may not happen (and if they do, will very possibly be the preserve of the less well-off in society) the push for ever more exotic vacation and second home spots will continue relentlessly.

Patagonia, in the south of Argentina and Chile, became so popular for reclusive celebrities (including George Soros, Sylvester Stallone and Ted Turner) in the 1990s that at one stage, a sixth of the region was said to be owned by 350 foreigners. Other way-out holiday spots for the future, as predicted by *The New York Times*, include Tibet, Tasmania (for nature trips), Cambodia (for the beaches) and Afghanistan (for skiing).

Could all this leisure not become an oppressive force in the near future? Are not the majority of future leisure trends strongly suggestive of a boom in little more than 'vegging out' and progressively atrophying mentally as we enjoy virtual sex over the internet, eat in ridiculous no-food restaurants and plan our next skiing trip to Afghanistan?

'If all the world were playing holidays,' the workaholic Shakespeare wrote in *Henry IV, Part One*, 'To sport would be as tedious as to work.' And that may be how life in the future seems as if it might be to us now. But go back further, to the Bible, in *Ecclesiasticus* and an interesting thought can be found suggesting that even thousands of years ago, the need for leisure was recognized.

'The wisdom of a learned man cometh by opportunity of leisure,' one passage of the ancient text reads. 'And he that hath little business shall become wise. How can he get wisdom that holdeth the plough, and that glorieth in the goad, that driveth oxen, and is occupied in their labours, and whose talk is of bullocks?'

The concept of ancient workaholics boring on about the latest trends in bullock management when they should be attending to matters of philosophy is alarmingly like a Monty Python sketch.

But it is heartening to think that, according to people of hundreds of generations ago, our increased leisure in the cash-rich, time-rich future could possibly lead to a new golden age of the mind.

Chapter 8

WHAT'S THE BIG IDEA?

Politics and Society

Keizo Obuchi, the Prime Minister of Japan, who died aged sixty-two in 2000, was a consummately dull man. Dreary in appearance, he was described as 'plodding', 'ordinary' and 'average', with no charisma or speaking ability. An American political analyst described him as having 'all the pizzazz of cold pizza'. His only spark of individualism was his habit of telephoning complete strangers out of the blue to ask their advice on policy matters. For all his dullness, he was held in great affection and esteem by the Japanese people for his modesty, gentleness and lack of pomposity. He weathered the endless cold pizza jokes with good humour.

In other countries too, both democratic and autocratic, the attractions of not living in exciting times seem to be increasingly prized. In Argentina, President Fernando de la Rua was thought to have won his 1999 election largely as the result of an advertising campaign which stressed how boring he was – a welcome change after the bombast, ostentation and daily drama of the outgoing Carlos Menem. Argentina is still congratulating itself on the growing dullness of its politics. In China, the likely successor to President Jiang Zemin, Hu Jintao, a fifty-six-year-old former hydroelectric engineer, is both politically conservative and, so it is said, more or less personality free.

In policy, equally, there is discernible all over the world a steady drive towards a dull, centrist, mixed economy, where people vote for parties whose real differences are only subtle, and the choices on offer are mainly of personality. This consensus, making us seem sometimes like zombies marching to some distant beat, is by no means a mere lazy, complacent settling for a status quo, either; a measure of progress and state intervention for social justice, howsoever feebly executed, is a standard part of the consensus package, whether in the United States or Peru, Sweden or Poland. Nominally left-wing governments in Britain, Germany and France now routinely pursue mildly right-wing policies at the same time as improved social justice.

The valuing of gently progressive political dullness takes the form in the Western countries of a virtual rejection of the day-to-day knockabout of confrontational politics. In Britain, the proportion of people voting in Parliamentary and local elections is on a steady decline, while in the United States, belief in government is at its lowest for forty years; in fact, there is a marked movement towards religion and voluntary community service as the preferred way of getting things done.

Historian Robert Fogel, director of the Center for Population Economics at the University of Chicago, believes America is in the grip of a great post-political awakening of religious fervour. According to opinion polls, two-thirds of Americans believe religion can solve most contemporary problems. And while only a quarter of college students admit to discussing politics, three-quarters regularly take part in voluntary philanthropic activities.

Part of the modern consensus-policy package is a degree of humanitarianism, along with a commitment to some kind of environmentalism. Another tendency is to want to replace national sovereignty with supra-national institutions. After a disastrous hesitation over the human catastrophe in Rwanda in the mid 1990s and a lack of resolve in standing up to the aggression of Yugoslavia, the blue berets of the United Nations are starting to be more respected than ever before. And in the peaceful majority of Europe, perhaps as a continuing legacy of the horror of two recent world wars, there is a distinct impetus towards unity.

Countries like China, Russia, India and Brazil, which once balked for a variety of reasons at Western commercialism, are now fully enthused members of the capitalist club. And even in the world's last Stalinist state, North Korea, capitalism is being toyed with. The talk of Pyongyang is of farmers soon being allowed to sell their produce at markets, mere discussion of which would have been considered treason until recently. The possibility of restaurants and shops being run as businesses is also being officially discussed.

Despite dips and swerves along the way, with Russia in particular frequently appearing to be on the brink of returning to authoritarian ways, the number of people worldwide living in stable, broadly free, broadly democratic, broadly capitalist, rather boring societies climbs year on year. The figures are extraordinarily impressive, and are paid scant attention by doommongers, who argue that fundamentalist and nationalist hordes are on the point of taking over the world, and by hair-splitters, who get too bogged down in definitions of what precisely constitutes freedom, democracy and so on to see clearly the amazing change that has come over the way humans organize themselves.

A hundred years ago, almost nobody could vote or have any say whatsoever in their own destiny. Today, according to a Washington study group called Freedom House, 88 of the world's 191 countries are 'free', with 2.4 billion people – 40 per cent of the population – able to vote and enjoy basic human rights. On top of that, there are another 53 countries and 1.6 billion people who have elections and civil liberties, although in societies such as Russia and Nigeria, these institutions are still too fragile and threatened by malign forces of corruption and privilege to be relied upon.

From a baseline of perhaps a hundredth of the world's population, two-thirds of humanity, therefore, is now free to a significant extent. In the remaining un-free parts of the world, still more significantly, perhaps, there is a longing among a large proportion of people to be free like the rest of us. If that longing were ever fulfilled, and if China in particular were to democratize, the proportion of people living under conditions of freedom would rise to seven-eighths.

As veterans, now, of the fickleness of futurology, we know that straight-line extrapolation of current trends is rarely on its own the most reliable way of predicting the likely future. A twentieth-century tendency towards democratic freedoms should not lead to the automatic assumption that we will become still freer in the twenty-first and twenty-second centuries. In predicting how things will develop, we need to extrapolate, then to factor in the inevitable surprises and what I earlier called 'knight's-move' thinking.

However, in politics and the organization of our societies, it may well be that we really are swooping forward towards universal freedom, because we have already tried the knight's moves and found them not to our taste.

Colonialism, theocracy, communism, autocracy, fascism and a dozen other isms and ocracies have served as diversions over the past several hundred years, yet we have emerged in the twenty-first century with just two truly big ideas to our name – democracy and capitalism.

Whether it is a cruel illusion or not will probably perplex students of politics for the foreseeable future, but most people in most cultures seem to feel that even if they are poor – in other words, even if their most fundamental human need, for food and shelter, has not been satisfactorily addressed – the freedom, democracy and capitalism package is still their best bet for improving their lot, rather than fundamentalism or violent revolution. Capitalism always appeared, especially to the political left, to be an historic phase which would inevitably be supplanted by another. But, as Anatole Kaletsky, associate editor and chief economist of the London *Times*, asserts: 'Now it's clear that it is the permanent equilibrium state of human society. Everything else was a sideshow.'

Voltaire wrote in the eighteenth century of the overriding need for human beings 'to look after our own garden'. And if, beyond the simple, universal human desire to live safely, quietly and comfortably, there is a single driving force behind this extraordinary, international, cross-cultural revolution towards a free, un-coerced life, it is individualism.

Whether or not it is possible, economically and environmentally, everyone in the world seems to want to run their own life, with their own plot of land, their own tractor, their own family and the freedom to follow their own culture and religion – or whatever local variants there might be on that model. (In the West, it might be, own apartment, own e-business, own partner of opposite or same sex, freedom to follow own hobby of collecting beer mats.)

Attempts to wean people away from this kind of nuclear family-based, low-level materialism have been notably unsuccessful. In the Soviet Union, even though millions who have actually become poorer in the post-communist era are said to be hankering both economically and psychologically for a return to a command economy, millions more seemed mysteriously to know, despite having no previous experience, exactly how to prosper in a capitalist economy the moment it opened for business. The way in which former Soviet citizens seemed to morph overnight into canny traders and business people echoed something a British conservative politician, Kenneth Baker, had said in Leningrad in 1988 as the USSR began to disintegrate and private enterprise started to be encouraged. 'This is going with the grain of human nature,' Baker enthused. 'It's not stroking the cat's fur the wrong way.'

It was not only in the communist bloc that full-blooded socialism (as opposed to cautiously socialist ideals like social security provision, which actually underpin much of the current Western policy consensus) ultimately found no takers. In Israel, for example, where the bulk of the country's current intelligentsia was raised and schooled under unbendingly socialist rules on *kibbutzim*, there has been a swift and overwhelming rejection of that collectivist lifestyle. Remaining *kibbutzim* are now peopled by a mix of elderly Israeli pioneers and idealistic young Germans and Scandinavians, while those born on collectives either live in cities working in academia or running hi-tech companies – or farm for themselves.

In the West, too, perhaps the only wide-scale movement towards broadly collectivist ideals – the caring, sharing hippies of the 1960s and 1970s – have tended to grow up into clever entrepreneurs like Ben and Jerry, the ice cream magnates. And even they finally sold out to a multinational.

Kenneth Baker spoke of human nature as the root of free enterprise, but for many, this will be less than satisfactory. What has really happened, it seems equally likely, is that people all over the world have simply settled for living as near to an American lifestyle as they can. And they have discovered that lifestyle not as a result of some universal *Little House on the Prairie* gene – but because they have spent a hundred years being exposed to the pleasures and pains of life in the United States via Hollywood and TV, music and literature – and come to the decision that they like it.

The fact that the twentieth century was an overwhelmingly American epoch, whether this was desirable or not, and whether or not the twenty-first will follow suit, exercises the minds of people all over the world, including in the USA, where there is a joint measure of pride and guilt on the issue.

The extent of the cultural takeover this new nation effected during the last century easily equals Roman domination during the period of history that state called its own. Just as the Roman Empire's enemies often wore togas, spoke Latin and enjoyed Roman inventions such as sewerage systems, even those today who are passionately opposed to American world domination wear its clothes, use its language and take advantage of its technologies. As one twenty-one-year-old woman obliged to attend an anti-American rally in Tehran, with bright colours showing beneath her black chador, told a *New York Times* reporter: 'It's a joke. How can you shout "Death to America!" when you're wearing blue jeans?'

Elsewhere, Islamic fundamentalists use US-designed laptops and log on to the American-developed internet in American English, while it's a fair bet that no known anti-American activist has yet been known to refuse to travel on a Boeing in preference to, say, a North Korean or Iraqi-built aircraft. Indeed, what is referred to today as globalization – the emergence of an international community which trades across national frontiers as if they did not exist, polices local wars in the name of humanitarianism and democracy, and ensures that shopping malls look (and are) the same from Guatemala City to Beijing – is effectively no more than Americanization disguised under a more palatable name.

American technology being so superior to the rest of the world's (as was Roman, and in the nineteenth century, British), it is easily forgotten how many of the benefits of globalization are wholly American.

The internet, for example, is enabling the most unlikely groups of people to improve their lives, very likely unaware that they are the beneficiaries of American industry. Farmers in Sri Lanka use the net to get up-to-date crop prices and negotiate better deals. A farming co-operative in Peru has reportedly quintupled its income by going online and selling to an inter-national export company. In Papua New Guinea, the internet is being used to spread folk knowledge on coming storms from village elders to a wider public. African villagers from Burkina Faso to Tanzania have expressed interest in setting up websites to sell traditional produce and artefacts abroad. And in India, villages without power are pioneering the use of solar-powered computers to access the internet and enable people to check on such matters as medicine stocks in distant health centres.

These may all seem rather gimmicky examples of American gizmos insidiously spreading the American way. Yet it is far from fanciful to say that every one of these technology-based advances also represents a severe blow to the ambition of authoritarian politicians.

Dictators fear and loathe their subjects having access to information. In the late Soviet Union, it was communications, particularly the fax machine and satellites, which helped bring about the fall of communism.

Just as it is quite realistic to argue that the internet (along with the likes of CNN) has put individuals in the ascendancy against tyranny, it is arguable that the onslaught of Silicon Valley technology has changed the very structuring of societies. The advent of railways and telegraphs, after all, facilitated the birth of nationalism in several countries in the nineteenth century, so it is not implausible that the internet might now be doing the same for internationalism.

Nations were once ranged against one another with their populaces stacked vertically, hierarchy on top, cannon fodder at the bottom. But now, there are the first signs of societies being organized horizontally. An environmentalist, a Talmudic scholar or a philatelist in Slovenia now has more in common with people of like mind in New Zealand than with people in his own country who don't share his interests.

Being able to discuss stamps or the trashing of GM crops with people around the world in what are sometimes described as 'transnational virtual communities' – or just buying cigars from abroad on the internet – may seem like relatively trivial freedoms. But they represent a fundamental advance of the individual over state boundaries and protectionist trade barriers.

The result of this process may be a whole new political structure, where overlapping governments oversee different areas of life instead of national governments ruling within state boundaries. Ultimately, in the coming century, it is said, the horizontalist, boundary-blind, new mindset will lead to traditional power politics being completely overridden by a global ideology based on human rights – and the final end of war as a result.

All this, optimists argue, and more, will come from the American invention of the internet – unless, as things work out, the dangerous little nationalisms festering away in areas like the Balkans do not fade away, but rise up to subsume the whirlwind of globalization. All that can, perhaps, be guaranteed is that national sovereignty and horizontalized globalization will be at loggerheads for much of the twenty-first century.

It continues to be argued, both in the United States and abroad, that the American century must soon come to a close. There are any number of good reasons for this to happen. Nuclear proliferation, wild economic instability, software catastrophes (possibly brought about by cyber-terrorism) all threaten to undermine even the most powerful country ever seen. The rot for America could come from its demographics. As the economies of Asia become more vigorous, America's population will be aging, and could yet be riven by linguistic clashes between Spanish- and English-speaking sectors of society.

And yet there are equally good reasons why American culture could hold sway as long or longer than did the Roman Empire. For one thing, the world refuses to abandon its America fixation. Both radical Americans and those in the rest of the world who are ideologically revolted by such American cultural imports as Rambo, gangsta rap and McDonalds (or for that matter, by the American insistence that women and gays should have equal rights in society) cast the Ugly American as the very devil of the modern world – 'The Great Satan', as the Ayatollah Khomeini's Iran described Uncle Sam.

Yet, frustratingly for anti-American voices, American culture, as exported worldwide under the banner of globalization, includes a continual and quite disarming critique of itself. For how does everyone from Barnsley to Burundi know so much about the underbelly of American culture, from the atrocities against native Americans to the urban poverty to the drive-by shootings to the crack houses to Watergate to My Lai to the bombing of Cambodia to the excesses of the CIA?

Well, certainly not from the propaganda put out by the news media of the former USSR, nor by modern enemies of the United States such as Saddam Hussein. Non-American journalists such as the Australian John Pilger or the Englishman William Shawcross have done their bit. But principally, we know about America's faults because American culture, from the Blues and Steinbeck to CNN and Hollywood, has for a century been turning it into prime-time entertainment.

America reminds us all about its own inadequacies, and yet somehow, this strengthens the attraction the culture holds for those who don't have it. Time and again, in both movie fantasy and White House reality, the highest institutions in the land are shown to be fatally flawed. And yet the USA continues to be far and away the top dog nation.

Additionally, unlike Rome or Great Britain, the United States and its culture has a special legitimacy in that it was in itself founded as an experiment in globalization. As far as the founding fathers were concerned, they had discovered a new planet. Coming from different parts of the old world, they were determined not to repeat the mistakes of that distant, left-behind life. They based the constitution of their new (albeit colonized) country on what were then ideas which seemed to have a global application, whether they came from Christianity or revolutionary French politics. America thus emerged, for all its manifest faults, as the first 'universalist' nation, designed to suit everyone and give everyone (slaves and women aside for the moment) the best possible deal.

It could well be that legacy which has continued for more than two hundred years to make American culture so attractive the world over – and may yet see it retain its supremacy for centuries to come. Not that F16s and Stealth Bombers will necessarily continue to police the world on humanitarian missions; internal US politics may force that baton to be handed over to Europe and others.

But American norms – the democracy/freedom/capitalism package which has so taken over the world – may continue to be pursued as a globalization which no longer flies the Stars and Stripes. For Pax Americana, read Pax Globalis. A revitalized United Nations or some successor to the UN may even be the agent of this new world order, with its increasingly faint, but still distinct, American accent.

But what about China? It is the widely-acknowledged heir apparent to global domination in the twenty-first century, yet it is actually far from clear how or why China should become the United States of this century. Yet apart from being the pre-eminent world power of 900 years ago and a very talented nation today with a very large number of people, most of the strikes seem to be rather against a Chinese century.

For a start, even though at the turn of the twentieth century Chinese novelists, smarting over a particularly bad time for their country, produced a clutch of books depicting a rampant, expansionist China taking over the world fifty or a hundred years hence, it has shown no sign whatsoever in recent decades of having any global pretensions. Typical of these books was one called *New Era*, published in 1908, which foresaw a massive technological battle successfully waged by China against the rest of the world in 1999, ending in China having bases in the Adriatic Sea. Economically and militarily still far behind America, however, modern China has never come closer to intervening in a distant foreign conflict than supplying technical help to the occasional embattled socialist regime.

Secondly, its language is extremely difficult to export and not wholly suitable for use on the internet. It would be inconceivable, however, for reasons of pride and linguistics, for China this late in the day to adopt the Western alphabet as Turkey successfully did in its modernization drive.

Thirdly, unless the People's Republic were to abandon its archaic political structure and, along with modernizing, embrace as its own the hugely successful Chinese diaspora outside China, it is unlikely that the country will in the near future move on from the massive task of pulling itself ahead, let alone impressing its culture on the non-Chinese world. While the rump of China remains politically opposed to that diaspora, it is disowning much of its brightest and best.

What, then, might we reliably expect of politics in the next century? Apart from a growing disillusion and lack of interest in domestic politics, the most probable drift, even if, as we might assume, nationalism continues to bubble up around the world, will be towards the operation of an increasingly viable and single-minded international community.

In many ways, we already have this, which was to an extent unthinkable even fifty years ago. In the realm of peacekeeping, it is now commonplace for armed forces to intervene in areas in which their nation has no obvious interest. In economics, the world – with the United States normally orchestrating matters – coalesces to assist any country in trouble.

Yet peacekeeping is itself a concept on trial. Part of the problem is keeping the peace after the headline-grabbing drama of 'good' wars, like that waged against Slobodan Milosevic in Yugoslavia in 1999. The British writer William Shawcross points out in his handbook on international peacekeeping, *Deliver Us From Evil*, that the capricious nature of television is often the initial impetus for peacekeeping efforts. Western television audiences want a quick fix to the spectre of children dying on their screens.

However, Shawcross writes, 'The television cry of "something must be done" can be irresponsible and fickle. Too many of the efforts I have described were forgotten as soon as victory was declared. Even Kosovo had virtually disappeared from newspapers and screens by October 1999. A commitment to peace is as important as a commitment to war, but is far more difficult to sustain.'

Nevertheless, the continuing presence of TV in international trouble spots, along with the new influence of the internet allowing oppressed people in one part of the world to address the public directly in another (as did anti-Milosevic Serbs during the Kosovo war), should ensure that international peacekeeping becomes an ever-stronger movement in the twenty-first century.

This is not to say that war will become obsolete, nipped in the bud wherever it threatens. But the late twentieth-century trend for conflict to be concentrated into isolated hotspots will, most experts concur, continue. 'For the foreseeable future, a third world war seems unlikely; there is no major ideological fracture severe enough to sustain it,' says James Mayall, director of the Centre for International Studies at the London School of Economics. 'War itself won't disappear, but it will continue to be regional or local, more often within the state than between states.'

The technology of war, one of the areas in which mankind has always been at its most inventive, will therefore continue to 'improve'. But here lies hidden one of the more curious potential surprises of the future.

In the past, increasing technology in warfare has always meant more soldiers and civilians dying. But with military technologists devising ever more smart weapons and robotic forms of warfare – from unmanned aircraft to mechanical infantrymen – the prospect of war without death looms.

We had a foretaste of this in the Kosovo campaign, when, for the first time in history, the victorious party suffered zero battlefield casualties. Cyber warfare, in which belligerents concentrate on crippling the enemy's computers, will doubtless become more common. The puzzle for the future may be, what if neither side in a war suffers casualties? How will conflicts be resolved? Will they go on endlessly, or until one side or the other runs out of toys? And will humans feel somehow 'cheated' if war can be waged bloodlessly on both sides, like a computer or a chess game?

Technology will also continue fundamentally to change the politics of the world in more peaceful arenas. One of the most important tasks for globalization to tackle is the woeful position of women in many developing countries. It is a given that educated women in all societies marry later, postpone having children and boost economic productivity. Pandit Nehru said that by educating its girls, a nation 'can revolutionize its economy and its society'.

The internet clearly has huge potential in education, both in helping women whom governments would otherwise not spend money on, and in giving educated women in traditional male-dominated societies the chance to use their skills. In Japan, some 85 per cent of women have university degrees, but end up in subservient jobs. The rise of the female Japanese internet entrepreneur is therefore one of the hot tips for the coming century.

The Muslim states present a subtly different paradox, meanwhile. There is discussion in Islam of what is being called the *cyberhijab* or cyber veil, to allow women who are not permitted to work with men or go out in public to see the world and work in it virtually on the internet. Saudi Arabia already has several service providers for women only. Whether these classic knight's-move developments represent an advance in the status of women, or a further entrenchment of their oppressed status, is a moot point.

What is not is that technology is again changing the most basic architecture of the world's population. Today, only 2 per cent of people have internet access, and already, social and political structures are shifting or preparing to shift. The political effect when 10, 20 and 50 per cent of the world goes online – and stays logged on twenty-four hours a day – can not fail to be immense, as great as the whole world becoming literate a second time round. The potential for making money on the net by selling directly to markets you never knew existed exacerbates that effect, by promising to help people get out of poverty by their own efforts.

Technology seems to be the most efficient mechanism invented since literacy to deliver that most subtle of human rights, self esteem. It is one thing to assist people by charity or humanitarian aid; but to give them the tools to take control of their own world has gigantic psychological implications.

Some megalomaniac may yet dream in the coming centuries of controlling the bodies and minds of a huge populace. But when there is no poverty, hatred and resentment for such a dictator to exploit because people are prospering in a world community by cultivating their own cyber garden, it is going to be an uphill struggle for the forces of evil.

'The far future to me is quite enlightened because there are no secrets any more,' says the science fiction writer, James Halperin, whose first novel, *The Truth Machine*, took the kind of innovative, lateral slant on the future which might well be needed to achieve any proper grasp of where we are heading.

Halperin believes humanity has flourished because we're better at deceit than all other species, but that our tendency to deceive each other now threatens our survival. He predicts the building of a machine that detects lies with 100 per cent accuracy. The machine puts lawyers out of work as crime is eliminated, forces political candidates to tell the truth, and thus reshapes humanity. 'You don't really need secrets once everybody has everything they need,' he argues.

But will man's aggressive nature not pose a problem to all this? Even when we all have what we need, won't the urge to acquire still more, and to cheat and lie our way to getting it, continue to underpin and undermine human affairs?

'I'm hoping,' says Halperin, 'that we can cure a lot of our psychological problems in time before we have the technology where there'll be billions of people each one of whom has the ability to wipe out all life on Earth. It's a tough one, but it's probably surmountable.'

Chapter 9

TO BOLDLY GO

Travel

When I asked him what the single greatest change in life during his seventy-five years had been, the Oxford social scientist Michael Argyle did not hesitate. 'It's the transport. Not the speed of it, so much as the sheer amount. When I was brought up in Nottingham in the 1930s, a couple of houses in my road had carriages, for which the horses were kept in stables, but most people didn't really go anywhere very often.'

We have moved in a short time from that static lifestyle to a point where transport is one of any modern government's major public policy preoccupations. Not only does everyone seem to want to be everywhere at short notice, but our determination to be in a state of perpetual motion may well have damaged irreparably the very planet we live on.

It has also backfired in terms of its own objectives; the sheer weight of numbers wanting to rush about our cities has meant that doing so is measurably slower now than it was at the height of the horse-and-carriage era.

In the air, while travel is cheaper than it has ever been, delays, air traffic bottlenecks and waiting about for baggage conveyors have conspired together to ensure that travel has not grown significantly faster in fifty years. Even the coming of the jet airliner has not speeded travel up as much as we like to imagine. The propeller aircraft of the 1950s, such as the British-built Britannia or the Super Constellation, were crossing the Atlantic in eight hours. With the two-hour check-in period most airlines now insist on, it takes almost the same time, or longer, today. The space between seats in economy on 1950s aircraft was additionally as generous as in business class now.

Even though there was little of it actually going on as late as the 1930s, transport has still been at the heart of futurological dreams for the past five hundred years. The ancient Greeks tried to fly. Leonardo da Vinci famously designed a helicopter. And remember from chapter one the 1661 forecast of the clergyman and philosopher of science, Joseph Glanvill, coming as it did just 120 years before the revolutionary invention of the hot air balloon: 'To them that come after us, it may be as ordinary to buy a pair of wings to fly to the remotest regions, as now a pair of boots to ride a journey . . . It may be that, some ages hence, a voyage to the Southern tracts, yea possibly to the Moon, will not be more strange than one to America.'

Outstanding feat of prediction (or guessing) as Glanvill's was, it is also significant that he got rather ahead of himself. Without wanting to appear carping, a voyage to the Moon continues to be very much more strange than a trip from Europe to America, and will probably remain so for hundreds of years yet.

For transport is also the area of futurology where imagination consistently runs far ahead of reality. Glanvill was imagining space flight 340 years ago, and it is still in its infancy, whereas as late as thirty years ago, only a handful of progressive thinkers had envisioned the internet – which has already had a far greater effect on the world.

Progress in transport always gives the misleading impression of being incredibly rapid, yet in reality, it is slower than in practically any other sphere. Arthur C. Clarke wasn't accurate in choosing 2001 as the date of his manned mission to Jupiter. And while visionaries speculate on us e-mailing ourselves to distant countries by means of quantum teleportation, the 'Beam me up, Scottie' technology of *Star Trek*, our one attempt at faster than sound civilian transport, Concorde, is a flying antique – and it is considered a feat of technological genius that we have succeeded in digging a tunnel under the twenty-two miles from France to Britain.

Even in the technologically humdrum matter of personal transport by car, we seem consistently to take leave of our senses when trying to predict the future. Broadly speaking, cars have not come very far in a hundred years. They look a little different, are cleaner and more reliable and have more electronic gadgets on board, but the design of a modern car would not astonish a mechanic from fifty or a hundred years ago. Our futurological predictions for personal transport, by contrast, appear to be on some kind of constant drug high.

It seems almost unbelievable, for instance, that the British inventor Sir Clive Sinclair was taken seriously and regarded as a pioneer in 1985, when he introduced his C5 electric tricycle, a low-slung plastic tub powered by a washing machine motor, which Sinclair – and much of the media – believed people would use on traffic-clogged city streets, where they could sit at the exact height of other vehicles' tail pipes. A commercial catastrophe, the C5 did find favour within a few months as a golf cart in dry regions of the world, such as the Gulf and Arizona, where damp did not interfere with its fickle electrics.

Having weathered similar opprobrium, the flying car dream is, unbelievably, still not *quite* dead, even if it is now the twenty-first century and the concept was apparently buried half a century ago. A Davis, California, company, Moller International, is having one – presumably final – stab at the idea with its eight-engine, three-computer model M400 Skycar, which is scheduled to have its maiden flight in 2000. Or maybe 2001. It was supposed to have been in 1999, but technical problems have inevitably dogged the project.

The vertical takeoff Skycar will initially cost $1m, and have a 900-mile range, flying at 350 miles per hour at heights of up to 30,000 feet. It carries two parachutes in case the engines fail. Dr Paul Moller, the Skycar's inventor, has spent £75m on the project – most of it from 400 private investors – and says he has taken 100 advance orders.

What of the car industry's more mainstream plans for the future? Well, they are most remarkable for looking just like future concept cars of 1950 did, as if we have somehow fallen in love with a universal concept of the motoring future which will always be dangled a few years ahead of the present time.

Generation after generation, body shapes continue to be 1930s-style Batmobile-streamlined. Steering wheels, we learn, will be replaced by joysticks, petrol engines by hydrogen-powered fuel cells. And there will be the ever-predictable on-board computer, which (a modern touch) will provide a constant stream of e-mails to the driver.

Although it is a fair bet that a car in 2100 will *still* be recognizable to Messrs Daimler and Benz, with a wheel in each corner and an engine, there are nevertheless some real developments making their way on to the market.

Satellite navigation systems, for example, have gone from being rudimentary to almost usable in a few years. The Michigan auto-technology firm Visteon is working on integrating voice recognition with route-finding, with the aim of producing a system where you dictate any address into the car's computer and wait for it to guide you to your destination – or somewhere that sounds reasonably similar, no doubt.

Several safety systems with a fair chance of seeing the light of day are also in development around the world. Computers which work with radar to anticipate collisions and apply the brakes for you are promised for some time soon, as are others which run tests on the driver to predict if he or she is likely to be accident-prone today. Many drivers will, of course, respond with some fear to the prospect of being put at the mercy of yet more software, when practically every program we use crashes on a daily or hourly basis.

Yet one of the land transport developments which even the more sanguine amongst experts predict really will happen is the automatic road, which drives the car for you. Professor David Newland, head of Cambridge University's Department of Engineering, for instance, is sceptical about the development of fuel cells – partly because the cheapness of gasoline in the United States acts as a disincentive to progress – and questions whether satellite navigation can deliver routing instructions which can genuinely be absorbed from behind the wheel. But he does predict a conceptual merging this century of the car with the train.

'The next big step in transport technology will be automated roads, regulating vehicles in convoys on motorways so that they're safer and can be packed closer together. You would just pay a toll, couple your car into an electronic convoy and sit back to enjoy the journey.'

Highways, Professor Newland says, 'will gradually become more like railways, with freight vehicles electronically coupled in trains running at relatively high speeds. At suitable intervals, they would uncouple to travel the remainder of the journey with their own driver. That's almost certainly going to happen. It would make better use of the roads and be safer, cheaper and greener, as well as making driving more pleasant for everyone. In Adelaide, there are already buses that run on an automated route for part of their journey.

'In fifty years' time,' Professor Newland concludes, 'driving your own car on a fast motorway, mixed up with lorries and passing at a closing speed of 150 mph within a few metres of people driving other vehicles in the opposite direction, will seem complete insanity.'

Our ideas about the future of flight have always been oddly more restrained than those about the more mundane business of driving. We have been in and out of our 'cities in the air' phase, where multi-level transport, with layer upon layer of flying machines, people strapped to body rocket belts and the like, was assumed to be the way ahead. Some futurology, particularly the more ironic kind, still likes to imagine city skies abuzz with mini-helicopters; the year 3000 cartoon *Futurama* shows 'New New York' swarming with such machines.

But practicalities such as the ridiculously high level of danger and the fact that helicopters are such capricious machines they have to spend several hours being serviced for every hour in the air, have, for the most part, consigned such ideas to fiction.

Like the flying car, however, the around-town flying taxi concept never quite lies down. There is a movement to reinvent the 1930s version of the helicopter, the autogiro, which was used by the military and to deliver mail. The autogiro has developed into the gyroplane, a hybrid small plane and helicopter, which can fly at 500 mph, stop in a short space, like a bird, and land vertically. The gyroplane is perfect for rapid, gridlock-hopping transport in urban areas – so long, obviously, as there aren't too many in the air at the same time. A company in Shanghai, China, is proposing to use 200 three-passenger mini-gyroplanes for an air taxi service, and it is not hard to imagine similar operators setting up in crowded cities like London and New York. Perhaps.

While the trend in commercial mass air transport is currently towards 'super jumbos' such as the European A-3XX Airbus, designed to carry 650 passengers on two decks, with shops, sleeping quarters and a gym, the return of 1960s-style ideas on super- and hypersonic aircraft cannot be ruled out.

Gargantuan fuel consumption aside, one of the factors which always held Concorde back was that it was hard to find many routes it could fly on at supersonic speed without breaking windows all along its route. The increasing global importance of the Pacific Rim area, however, could herald a comeback for ultra-fast mass transport.

'What people hate about long-haul flights is boredom, lack of space, dehydration and jet lag,' says John E. Ffowcs Williams, Rank Professor of Engineering at Cambridge. 'The flight from Seattle to Australia currently takes twelve hours. Just think how delighted everyone would be if you could cut that to three. There would certainly be a market. So much so that I could imagine the Pacific crossing driving the development of supersonic passenger aircraft in a way the Atlantic never has.'

Perhaps as a result of this new perspective, NASA has allocated two billion dollars to develop a 300-passenger aircraft capable of operating over 5,000 miles at Mach 2.4 – fast enough for a two-and-a-half-hour Seattle to Tokyo flight time. One of the engineering avenues the research is exploring is the hypersonic ramjet, a radical new type of jet power which uses the aeroplane's forward speed to compress air and use this to power the jet further. Suborbital Mach 10 loops made by such airliners, with their flattened, knife-shaped fuselages, could shorten intercontinental flight times to less than an hour.

Development of sea travel will not be abandoned in coming centuries. On the contrary, cruise liners are becoming more gigantic and luxurious every year. And the new form of residential cruise ship mentioned in chapter two, where passengers install their own furniture on board and establish the high seas as their permanent address, is about to start catering for well-heeled people who like the idea of continually going round the world – but would really prefer the world to go continually round them.

Such liners will typically spend the majority of the year moored for days or weeks at a time off the world's major travel destinations, preferably at the time of a big event such as Carnival in Rio, the Cannes Film Festival or the Americas Cup. And as befits communities based on home ownership rather than just renting cabins for the occasional vacation, there will even be a form of small-town democracy on board, with residents voting on where the ship goes next.

There is smart money, too, on undersea leisure travel eventually becoming popular. It is difficult, indeed, to imagine luxury submarines not eventually taking rich and non-phobic adventurers on deep-dive cruises to explore the mysterious, dark regions miles beneath the sea. Perhaps as a precursor to such developments, the first – and, to date, only – undersea hotel was opened off the Florida coast in 1986. Jules' Undersea Lodge, near Oceanside Key Largo, is located at the bottom of a natural mangrove lagoon in thirty feet of water. It has bedrooms and a dining room and advertises itself as the ultimate romantic honeymoon destination – so long as guests are happy to scuba dive their way to reception. There is no other way in. NASA has used the lodge as an economical way of researching what might be required for extended space travel.

Faster, bigger, more automated, more luxurious . . . conventional wisdom suggests that all we need do to know about the transport of the future is to imagine a simple, linear improvement of what already exists, without any conceptual shift at all.

To be fair, in the case of some types of transport, especially railways, it is hard to picture what conceptual shift there could be beyond making trains faster, smoother and more reliable; the greatest advance in rail travel would be the one which evades most governments – to make it cheaper than going by car to compensate for the intrinsic inconvenience, that a train cannot take you and your baggage from door to door.

But to find out what route the journeys of the future might really take, it may well be necessary, as in the old joke, not to be starting from here. Might it not be that transport in the future, when we have practically all the information and sensory experience we could need piped into our lives via the internet, will become slower, and something we will be inclined to do less of?

Having artificial travel (as well as pleasure, education and sex) beamed into our homes rather than having to go out and physically convey the lumps of meat we call ourselves over tens and hundreds of miles sounds like dehumanization, as if we will turn into what British science writer Paul McCauley describes in the technology magazine *T3* as '150 kilo snack-fed grubs locked in individual cells of honeycomb cities, as coddled as babies and as omnipotent as gods'.

Yet an activity like hunting has changed from being a daily, life-and-death necessity to a leisure pursuit, a frippery, in a few hundred years. We do not feel less human for having supermarkets do our hunting for us, but to anybody from a few generations ago, it would seem that we have already turned into McCauley's spoilt, fat grubs. Perhaps travel could go the same way, and be regarded as just as valuable when delivered virtually, online, as when it is done in the flesh?

Travel and transport will be unnecessary in science fiction writer James Halperin's future. Virtual reality is, for him, an altogether more likely prospect. 'If we can control the environment completely, location won't matter any more,' argues Halperin. 'Some people will, perhaps, want the real thing in terms of travel. But I'm not sure I know why; once you can interface with somebody as if you're having a conversation in the same room, or once you can explore a place as if you're there, what's the difference? But some people will just want to say they did it.'

'Air travel has lost its glamour,' observes Professor Ffowcs Williams of Cambridge. 'There are few exotic, rare places any more – cheap travel is ruining all that – and the communications revolution now makes it practical and nicer working from home. So I can imagine aircraft being used less in the future. At the moment, capacity is still rising, but in twenty years, it will be falling unless the airlines get us about faster or make it more fun, like the early 747s, which had a piano bar on the upper deck.'

Concern about tourist hotspots being swamped by crowds which destroy the very things they have come to see can, curiously, be traced back nearly two centuries. A cartoon by George Cruikshank in the London *Comic Almanack* in 1842 depicted 'Egypt in the future'. The only people for tourists to see are other tourists, or Egyptians in what we would now call the tourist industry – i.e., selling things to them. The Pyramids themselves are bedecked with advertising.

However, even if virtuality can supply most of the bodily thrills we need, holidays included, unless nanotechnology fulfils its somewhat unlikely promise of making it possible to download from the internet the pattern for anything from a lettuce to a lawnmower and build it in our homes, atom-by-atom from a 'toner' of raw materials, freight transport is going to become a bigger and bigger business. This is not just because there will be more people to consume freight. As trade, in both the West and the Third World, intensifies and there is a general trend towards greater prosperity, proportionately even more goods will be required than is now the case.

Transporting vast tonnages by 747 freighter, as we now do, is clearly wasteful, expensive, unecological – and for many goods, unnecessarily speedy. Ships, on the other hand, are still painfully slow, and involve vast fleets of container trucks to offload cargoes at port and drive them noisily and disruptively to their final destination. Could the next century, then, finally see the re-emergence of that stalled, but wonderfully logical, relatively fast and passably green technology, the airship?

Airships seem to have been threatening a comeback for decades without result, but several projects currently promise to relaunch the technology, which seemed to have died with the crash of the R101 in France in 1930 and the *Hindenburg* in New Jersey in 1937. A German company, CargoLifterAG, was floated on the Frankfurt Stock Exchange in May 2000 and raised £75m. Working from an old Soviet airbase outside Berlin, CargoLifter plans to operate, starting in 2003, a fleet of 200 helium-filled airships each 260 metres long, with a range of 6,000 miles – and able to transport goods the weight of ten fully-loaded trucks.

A British company, Airship Technologies, is meanwhile planning an even more spectacular venture in conditions of some secrecy from the R101's old hangar near Bedford, England. The company's *Skycat* ship, according to British documentary maker Tony Edwards, who has filmed its development for a Discovery Channel series to be shown in 2001, defies conventionality by being in the form of a flying wing rather than the traditional cigar shape CargoLifter is sticking to.

Skycat will be three times the size of a large sports stadium, with a length of 1,000 feet, but because of its shape, its designers believe, will be more stable in high winds than the German competition. *Skycat*'s payload will be an estimated 1,000 tons, and it also has the benefit of not needing to be tethered to a tower for loading and unloading – which, in CargoLifter's case, will probably have to be done by crane.

Skycat has instead a patented soft landing system, by which the craft sticks itself to flat ground by vacuum, and cargo simply drives off. The Pentagon is among parties looking with interest at the project. Military minds are attracted by the possibility of being able to move 1,000 tons of troops and material in one, non-stop flight across the world, from the heart of the USA to any trouble spot – albeit at a stately 80 mph or so.

Airships are one lateral-thinking transport solution; at the other end of the scale is the wackier question of quantum teleportation. The theory behind teleportation is unimaginably complex, dealing as it does with the common-sense-defying paradoxes of quantum mechanics, the theory of how matter behaves at the level of individual atoms and smaller units still.

The idea of an object, be it a particle of light or a human body, travelling instantaneously, in no time at all, and reproducing itself perfectly elsewhere by what Einstein called 'spooky action at a distance', seems to defy the great man's own Special Theory of Relativity. Einstein predicted that travel at beyond the speed of light is impossible, because our mass will approach infinity as we try to go the distance between 99.9999 per cent light speed to the full 100 per cent.

Teleportation additionally appears to contradict Heisenberg's Uncertainly Principle, which states that it is not possible to know both the position and the velocity of an electron. *Star Trek*'s creators thoughtfully included a 'Heisenberg Compensator' (of unknown design) in their teleportation machine. However, in 1993 a team of physicists from IBM, the University of Montreal, the Technion in Israel and Williams College in Massachusetts, demonstrated that quantum theory could circumvent, but not violate, Heisenberg's principle – and that teleportation was therefore possible.

Within four years, the Austrian physics professor Anton Zeilinger and a team then at the University of Innsbruck reported in *Nature* that they had successfully teleported a particle of light, or more accurately, a quantum *state*, from one side of a laboratory to another, and a whole new raft of teleportation speculation was launched. Already, there is talk of one day teleporting larger items, perhaps even something as massive as a whole virus.

But teleporting whole bodies, if it is ever achievable, is most probably centuries away; it would require instantly scanning the position and identity of every atom in the body, coding the information for transmission and unscrambling it at the other end – an enterprise millions of times bigger than the Human Genome Project. And after all that, a fundamental question would remain unanswered until the first brave volunteer stepped into the teleportation chamber.

Even if a perfect replica of a person – not a 'fax', but an object indistinguishable in every sense – could be made to appear in another place, would the person's mind be present in the replica? Or would it have been left behind in the (destroyed) original?

Another transport technology of sorts which the expert consensus has it will not be with us any time soon is time travel. Aside from Professor Stephen Hawking's apparently insurmountable objection – that if time travel ever has or will exist, we would surely have seen some travellers from the future by now – the technology required to put such a speculative theory to the test could only realistically be available to a super civilization – and not, one suspects, to one which at any time in its recent past rated digging a hole all the way from Calais to Dover as a noteworthy achievement.

Hawking also believes, as part of what he calls his chronology protection conjecture, that the laws of physics, quantum or otherwise, will always conspire to prevent time travel – even though he fervently maintains at the same time that the paradoxes of quantum mechanics, a science barely understood even by many scientists, 'will seem as just common sense to our children's children'.

Of course, in a sense, 'natural' time travel exists already. In accordance with Einstein, astrophysicist Professor Richard Gott of Princeton has calculated that the astronaut Story Musgrave, who spent 53 days in Earth orbit repairing the Hubble telescope, returned over a thousandth of a second younger than he would had he not gone on the mission. An astronaut spending 30 (exceedingly hot and uncomfortable) years on Mercury, which has a smaller orbit than ours and circles the Sun faster, would have a more calculable youth bonus of 22 seconds, while by simply travelling at 99.995 per cent of the speed of light to a point 500 light years away and returning, you could visit the Earth in 1,000 years' time, while having aged just 10 years yourself.

As far as building a time machine goes, Professor Gott explains that it would require constructing a 'wormhole', a tunnel that acts as a shortcut between distant sections of the space-time continuum when the existence of a very large and heavy object causes space and time to curve.

Building such a wormhole would be no problem to engineers, so long as they were up to making an object 600 million miles in circumference, with a mass 200 million times that of the Sun. Alternatively, they could exploit Gott's own contribution to astrophysics, known as string theory, which would merely require the harnessing of half the mass-energy of a whole galaxy in order to venture back in time by a year.

So it seems that there is, will be and always has been a gloomy outlook for time travel. However, as for the old time travel conundrum of what would happen if someone went back and killed their own grandfather, thereby making their own existence an impossible paradox, theoretical physicists do have an answer.

The Oxford University physicist David Deutsch, author of *The Fabric of Reality*, one of the best guidebooks to the baffling quantum world, is a leading proponent of the theory that if you did kill your own ancestor, space-time would obligingly open up a parallel universe in which you did not exist, and that there can be an infinite number of parallel universes which trundle on without interfering with one another. Some of us may, of course, conclude from such staggering ideas only that there is, indeed, at least one parallel universe, and it is where theoretical physicists live. But it must be noted that this is mainstream physics, not the paranormal; the equations say so.

I have left the question of space until the end of this discussion on the future of travel, because it is entirely unclear whether, beyond having quite an active life on space stations in the immediate, orbiting vicinity of Earth, humankind will ever travel in person to other planets, or beyond that, into interstellar space.

This is not to say we will not explore space; our record in the short time since Sputnik 1 was launched in 1957 has been extremely impressive and continues to be, but more in the unmanned probes department than in the more sideshow-ish business of sending human beings into space.

Glamorous though manned space travel is, it is becoming

increasingly evident with hindsight that it is not hugely necessary, at least at this stage in the exploratory process. Its agenda is
even arguably dictated more by showbusiness than by the needs
of science. Fun as it was to see astronauts on the Moon playing
golf and saluting the Stars and Stripes, there was probably not a
lot they were personally able to do that could not have been done
by robotic probes in that flurry of manned space activity so
many decades ago.

Today, and in the next few years, unmanned spaceships
bristling with electronics are heading out, or shortly will be,
towards Pluto, Neptune, Saturn, Jupiter and its moons, Venus
and Mars. Some will orbit and relay back data; others will have
more ambitious missions. The Europa Ocean Observer will in
2007 be seeking out a suspected water lake under ice on that
intriguingly Earth-like moon of Jupiter. A series of Mars Sample
Return Missions will start bringing back rocks from Mars in
2008.

Travel, robotic or manned, *beyond* our own solar system is
highly unlikely in the coming century, however. Scientists are
only now puzzling out the problems of sending some kind of
probe to our nearest star, Alpha Centuri. A little depressingly for
Star Trek or *Red Dwarf* fans, the very edge of what is considered
possible for the twenty-second century, but not the twenty-first,
is represented by a proposed ship called *Starwisp* and designed
by Robert L. Forward, an independent aeronautical engineer
formerly with Hughes Aircraft.

Starwisp is a giant, round sail of fine wire mesh, a kilometre in
diameter, and carrying 100 billion microprocessors to act as
camera, brain and transmitter. This incredibly delicate structure,
which weighs just half an ounce, would be powered by blasts of
microwaves across space from a 10-billion-watt solar satellite.
About ten watts of that power would reach the huge, ungainly-
looking fishnet, but that would be enough to push it along at a
fifth of the speed of light. It would reach its destination in
twenty-one years, and its signals would take four years to travel
back to Earth.

Lightness in proportion to size is always going to be essential for interstellar spacecraft, which are hampered by the problem that the bigger they are, the more fuel they require, and the still bigger they consequently have to be to carry that extra fuel. A scaled up *Starwisp*, carrying passengers, is nevertheless conceivable, but would need to be a million times bigger. It might be possible five hundred or a thousand years after the first *Starwisp*.

Travel into Earth orbit is almost certainly the immediate future for the bulk of manned space travel this century. By 2006, the troubled white elephant International Space Station should be operational. Additionally, a variety of private companies are trying to get the first space tourists into orbit by 2005 on independently designed and launched shuttle craft. They are highly unlikely to be successful, although Hilton and Budget both have theoretical plans to open orbiting hotels.

While NASA has plans drawn up for VentureStar, the successor to its 1970s Space Shuttle, the hope for truly affordable travel into Earth orbit may just possibly rest with a concept invented by the Russian Y.N. Artsunatov in 1960 and taken seriously by Arthur C. Clarke, who first conceived of communications satellites in 1946.

The idea is for space stations and hotels to be connected to Earth by an elevator. Solar powered, such contraptions could cut to a thousandth of its present figure the vast cost of launching stuff into orbit. Hundreds of miles long, they would require some material stronger than we now have – but not outlandishly so – and could well by the twenty-second century have passengers riding up in gondolas, first to an interim space platform 160 miles above Earth, then on a second, more slender elevator cable for another, easier 600 miles in the weightlessness of space.

The ultimate aim of manned space travel beyond Earth orbit will be to investigate the possibility of colonizing Mars. But whereas the 1960s and 1970s Moon landings had more to do with showing off to the Soviets than with any real human need, there is a growing belief that we will actually need Mars sooner than we might imagine. Even if ecological disaster doesn't render Earth uninhabitable, a spin-off of ever-better medical care will be a population which increases relentlessly whatever happens to birth rates, because so few people will die.

The technology for travelling to Mars is fairly unremarkable – not much more than an enhanced version of the hardware which got Apollo to the Moon. All NASA really needs is the political will and the funds. 'There will be people on Mars long before the end of the twenty-first century. It's inevitable. It's irresistible,' insists Malcolm Walter, a palaeobiologist at the University of Sydney involved with NASA's continuing search for life on the planet. 'It might happen before 2020. It could happen by 2011. Mars is our next frontier.'

And according to Princeton's Richard Gott, too, it urgently needs to be. He is convinced we will improve our chances of survival by pressing ahead with colonization, but worries that we will abandon the quest for the new frontier, 'as the Chinese did in the fifteenth century. They explored Africa, came back with a giraffe that everybody wondered at, and then they just quit.' The Egyptians showed the same lack of resolve, Gott argues, with their pyramid building, the golden age of which lasted only a century, after which they built increasingly crummy ones. 'So there's a danger that we'll end up stuck on the Earth,' Gott says.

'Should something bad take us by surprise in 150,000 years – an epidemic or something like that – we'd be saying, "We should have colonized back in the twenty-first century, while we had the chance," ' Gott told the *New Yorker* magazine. 'I'm not saying Mars isn't dangerous, but colonizing Mars would increase our survival prospects by giving us two chances in the casino of life.'

And beyond colonizing Mars? 'The big question as early as the twenty-second century,' James Halperin believes, 'will probably be what do we do with Jupiter, which contains most of the matter in the Solar System that isn't at a million degrees. By dismantling Jupiter, we could build a thousand planet Earths in our orbit, with our climate – and accommodate 100 trillion people.'

Chapter 10

MYSTIC MEGABYTE

Future Present

The twentieth century saw two world wars, several instances of genocide, a great depression, a global plague and an international ideological schism that came within hours of ending the world. After all this, we have emerged into a twenty-first century in which a combination of science and global capitalism, loosely underpinned by an ethical consensus forged from the remains of socialism and the most attractive parts of the major religions, offers the best chance we have ever had to live happily, healthily and prosperously in an almost war-free world.

In spite of all these reasons to be cheerful, most of us think the past was better than the future will be. The present, we tend to believe, is brutal, ignorant, dirty, polluted, unstable and dangerous. Madness and crime are flourishing everywhere. Hordes of fanatics around the world, from the Taliban to American Christian fundamentalists, are being encouraged by shady movements to give full rein to their racism, superstition and fears.

The Western intelligentsia, who should be helping them through education and example to live a more enlightened, constructive, humane life, are consumed instead with their own greed, egomania and pursuit of a happiness which is elusive because it doesn't exist – or certainly not to the exacting standards they demand. In between seeing shrinks to sort out their boundless self-obsession, they cynically make money feeding the uneducated masses not enlightenment to counteract the evil of fundamentalism – but junk food, junk products and junk news about minor actors and minor royals.

If this depressing picture is not bad enough, isn't it obvious, we assume, that the future will be the same, but worse?

All this gloom, of course, ignores the fact that the past, even thirty years ago, was relatively horrible. For every modern unpleasantness – pollution, crime, ignorance, crassness, un-couthness – it is possible to come up with something as bad, or generally worse, from any one of the 'golden eras' of the past. Crime in a city like London may be an annoyance, but someone from Dickens's or Hogarth's day, or even 1930s policemen who had to patrol the city's darker quarters in large groups for fear of razor gangs, would regard London in 2000 as Shangri La.

A miserabalist view of the present similarly glosses over the fact that pessimism about the future has a disastrous past. Doom-mongers are invariably wrong, from Nostradamus to modern seers like Paul Ehrlich, whose 'population bomb' never went off, to Yale professor Paul Kennedy, who predicted in a 1987 book, *The Rise and Fall of the Great Powers*, that America was about to collapse and take the free world with it (when it was the Soviet bloc which was months from collapse) to en-vironmental activist Jeremy Rifkin, who forecast in a 1995 book, *The End of Work*, that computers would cause cata-strophic mass unemployment – only for them to become the biggest job creator since the industrial revolution.

The only accurate thing most pessimistic futurologists get right is their book sales forecasts, which always seem to fulfil their targets. Being downcast about the future appears to address a basic human need. We have long derided optimism, since the credulous Dr Pangloss ('all is for the best in the best of all possible worlds') was the butt of Voltaire's satire, and before.

I suspect there are three reasons for the popularity of this negativity. One is a kind of subliminal superstition; if we say things are going to be awful, who knows, might they not work out slightly better? The second is intellectual snootiness. It is wiser if you want to appear intelligent to expound on what things aren't rather than what they are. To a large extent, this is a good thing. Scepticism is a proper and respectable academic mode – we don't want our intellectuals to be too naïve – but it is also utterly safe and unimaginative.

The third reason we – and I include in that 'we' educated people all over the world – cleave to pessimistic prediction is that optimism seems somehow too shiny and pearly-toothed and American for educated people to be doing with. It is the mindset which brought us the 1950s 'Duck and Cover' atomic bomb protection advertisement, which advised Americans that everything would be OK in the event of atomic attack if they remembered to hide under a table. It is the smiley pseudo-philosophy of Disney and 'Have a nice day'.

Nevertheless, there are signs that optimism is now moving upmarket, initially – of course – in the United States, but across the world too. First, in 1989, came the dramatic assertion of Francis Fukuyama, a third-generation Japanese-American academic then working in the State Department, that liberal democracy marked civilization's terminal velocity – that no better form of society could exist, and history had, in effect, plateaued.

Fukuyama was surprisingly hailed by many academics across the world as having, with his 1989 book, *The End of History*, made probably the most important contribution to the philosophy of history since Marx and his communist successors. Now the professor of public policy at George Mason University in Virginia, Fukuyama became the instant icon of a new, optimistic modernity and as befits a man so up-to-date he gave up his hobby of making reproduction antique furniture from wood and now constructs ever more elaborate pieces virtually, by computer.

Hot on Fukuyama's heels, a 1996 article in *Wired* magazine by Peter Schwartz, a San Francisco futurologist, added to the revival of optimism – or, as some would say, complacency – by arguing again that we've never had it so good – and that the future was going to be even better, thanks to an economic Indian summer he called 'the Long Boom'. Schwartz is the German-born child of Auschwitz survivors. He became a highly successful oil industry analyst for Shell in London in the 1980s, and has now gone on to become a hero of corporate and middle-class America, counting Alan Greenspan and General Colin Powell among his fans. His article grew into a 1999 book, *The Long Boom – A Vision for the Coming Age of Prosperity*, co-written with Peter Leyden (a technology, economics, and political journalist), and Joel Hyatt (a Stanford professor of entrepreneurship). In 2001, Schwartz's upbeat vision will serve as the background to a Steven Spielberg film, *Minority Report*, starring Tom Cruise and Sir Ian McKellen. The film will, apparently, be set in a clean, bright, prosperous 2080, in which the police can pre-empt 99.9 per cent of crime by predicting it before it happens.

With its belief in the melting away of borders as cross-cultural ties on the internet supersede outmoded concepts of nationality, Fukuyama's optimism and the super-optimism of the Long Boom are eerily reminiscent of no less (or more) than John Lennon's sweetly naïve aspirations in his song, 'Imagine', albeit with modern groovy capitalism having replaced Lennon's vaguely communistic idealism.

The Long Boom also seems uncannily like an oversized, bumper helping of the arrogance of the present. How seduced *are* these people, I occasionally wonder, one an immigrant, the other of immigrant stock, by the goodies and the easy living of late twentieth-century American life?

Yet, when people ask what conclusions I have come to after a year or so of total immersion in futurology, I have to say I have found the modern American confidence one of the more convincing models of the future. This is not necessarily because it says something we desperately want to hear. Indeed, given the fierce anti-optimism of the post-war, first-generation university-educated middle class – a cynicism to which I generally tend to subscribe – Fukuyama's and Schwartz's world picture is quite a dissident one.

But more importantly, even if it *is* a bit too shiny and born-again for European palates, it still accords with the real lesson of futurology. This, as I see it, is not that things usually turn out slightly worse than the best forecast – but that they invariably turn out considerably better than the worst. Dystopic predictions – the opposite of utopian – practically always look hysterical in retrospect, while optimistic futurology frequently seems rather understated when we view it with hindsight.

The environment seems to me – if step out timidly into the moving traffic of futurology I must – to be a prime candidate for illustrating this principle. There will, of course, be natural disasters in the future. A major Tokyo or California earthquake, for example, is always possible, if not inevitable as the doom industry insists. But, given the woeful accuracy of past environmental prediction, I suspect we flatter ourselves in believing a killer asteroid, straight out of Hollywood central casting, will flatten us.

As for global warming, my suspicion is that the climate will get warmer, but that sea levels will not rise very dramatically. The intellectual argument over whether the warming and the loss of some land in countries like Bangladesh is feckless twentieth-century mankind's fault or a natural cycle in operation, will still be unresolved a hundred years from now. Some of the then very elderly global warming theorists will be congratulating themselves over having got it so right; the ancients who scoffed at the concept in 2000 will still be scoffing in 2100.

However, I believe the environment, in the sense of our cities and countryside, will improve vastly over a couple of hundred years. The air will be extremely clean and bright, the streets sparkling. The carefree way in which we currently pump the atmosphere, the soil and our food full of pollutants of unknown effect will be regarded as repellently outdated. People will see our era's essential filthiness as every bit as crude and incomprehensible as medieval open sewers in the streets seem to us today. Becoming routinely clean and green will be as fundamental a change as was the adoption of basic hygiene in the nineteenth century, which was seen as a deeply eccentric concept until Joseph Lister, around 1865, convinced a deeply sceptical – often openly scornful – medical profession that germs 0.001mm in diameter existed in the environment and caused septicaemia. (Before then, the unquestioned received wisdom was that inflammation somehow produced microbes.)

What do I imagine people will be like a hundred or five hundred years hence? I believe, although this is more hope than conviction, that the middle classing of the masses, which will come about by the simple means of making, mostly by education, an increasing proportion of the population more comfortable and better off, will lead to a higher average level of education, more humane attitudes and a diminution of crime and anti-social behaviour, as the broad mass of people begin to feel they have a real stake in humanity's future.

Many of our current problems seem to stem from individuals and groups of people in all countries feeling they have little in common with those who have their own homes and land, a car, and children who are hoping to do better than their parents. In Britain, Europe and the United States such alienated people turn to drugs, crime and public and private aggression. In a country like Pakistan, for example, disappointed by the corruption and nepotism of a new democracy, they turn to the vicious but honest-seeming certainties peddled by the Taliban, whose power there is rising ominously.

Seeing one's salvation as the handing over of power and responsibility to such as the Taliban seems to be as futile as taking out a loan to get out of debt, but that is what poverty and ignorance does – and it can be safely assumed that when people become wealthier and more educated they will be less prone to be tempted by brute fundamentalism. The same probably applies in Kansas, where the bizarre political power of those who deny evolution continues to grow. Such superstition-mongers will not prosper in a richer, better schooled world.

However, even if the creation of an ever more enormous global middle class is not a myth, as it might, of course, turn out to be, there is still going to be a less well-off layer in all societies.

My main fear for the future is that the rich/poor polarization which already exists could become much worse, with a state of near war in some places between haves and have nots. In Los Angeles, private security is rivalling computers as a growth industry. There are war zones in the city where AK47s can leave half a dozen people dead in a weekend. This is not perceived as being a public order problem of great importance, though. The police are said to code drug dealer on drug dealer killings as NHI – no humans involved.

It is extremely difficult to envisage a way out of such a corrosive state of affairs. Decriminalizing drugs may be a pragmatic answer. It does seem absurd that almost every member of every legislature in every democratic country has probably used drugs in his or her youth, yet they conspire to keep them illegal, and thus hand over the lucrative monopoly for their commercial supply to 'non-human' criminal elements. It sometimes seems that while drug dealers and addicts feed off one another, they are seen by governments as performing a useful self-culling exercise. What damaging mischief, after all, might criminals get up to if they weren't in the drug trade?

The United States may eventually give gun control a chance in the battle to curb crime, but every schoolyard shooting perversely seems to increase the National Rifle Association's membership, as a flight into fear by well-to-do people gathers momentum. At a guess – thinking, for instance, of how racism tends to be worse amongst people who don't actually live alongside other races – many of these Americans taking up arms in fear of crime will never need to use them. But the psychological damage and brutalization, which runs so counter to many of the optimistic trends in American society, will transmit culturally down through generations. As a result, private individuals will probably still be touting lethal weapons in 2100.

Crime will be at the centre of the twenty-first-century debate over genetic engineering. The genetic determinist proponents of the technology will insist they can edit crime out of the human genome. The anti-genetic engineering lobby will be outraged. Crime is at root a survival mechanism for the poor, they will argue. If the bio-engineers can really remove it from people's make-up, they will simultaneously block out important survival abilities. But the very idea that anti-social behaviour can be genetic in origin, other than in cases of mental illness, will anyway be the cause of derision, and very likely of political direct action to dwarf today's anti-GM crops activism. There will be a great deal of discussion about Australia, the only country with a substantial proportion of its population descended from convicted felons. Why are there not a disproportionate number of Australians genetically predisposed to crime, the political left will demand? Research groups sponsored by the political right will miraculously discover that there *are*.

While growing prosperity will, genetic engineering aside, cause crime to decline, there will continue to be increases in other forms of unpleasant behaviour. Most people in economically booming Britain, for example, feel they have noticed a marked increase over the last few decades in general rudeness, public misbehaviour, drunkenness and vulgarity. Even in peaceful Denmark, where there are virtually no social problems or poverty, there is said to be a spiralling level of road rage. Some will put this crude behaviour down to alcohol, others to liberal education and the breakdown of the family. As prosperity increases, there will be demands in many countries for stricter controls on drinking.

But those who believe public violence and misbehaviour are not a function of harmless social drinking, but of family instability and lack of education, will point to the example of Ireland, where people drink stupendous amounts but, well educated by the state and disciplined by a firm family structure, behave generally very well.

As with crime, a huge genetic engineering debate will also rage around homosexuality. The intelligentsia around the world will be deeply troubled later this century by the prospect of the standardization of human beings by genetic engineering, and the loss of original thinking that might entail. Legislation will prevent attempts at a range of genetic engineering, especially where it is designed to 'repair' divergent sexuality or prevent 'gay babies' from being born.

However, undeterred, parents will seek out genetic engineering clinics prepared to ignore the law and offer anti-homosexuality gene therapy. There will be much-publicized legal cases when these treatments 'go wrong' and parents attempt to sue clinics over their gay children, even though the original treatment was done illegally. The children may additionally, in some cases, sue their parents for having had the treatment in the first place. Needless to say, law will continue to be a lucrative profession.

Spirituality will almost certainly be a growing trend in coming decades and centuries, partly as a reaction to the ever more automated, robotic world, and partly as a response to people having a little too much time on their hands and finding a window in their schedule for a spot of self-fulfilment. Already it is routine for the most unlikely people, such as soap stars, corrupt politicians, wealthy socialites, publicly to declare an interest in 'developing their spiritual side'. To the frustration of priests and purveyors of organized religion, however, the growing interest in religion will increasingly be an eclectic, pic 'n' mix spirituality, which will seem to them to be more style than substance.

I have come across plenty of serious scientists, too, who attest to being intrigued by religion, referring, for example, to 'The Creator' just when you weren't expecting them to. Arthur C. Clarke is open about his interest in spirituality. He told Stanley Kubrick, director of *2001*, that it was 'the first ten million dollar religious movie'. 'I don't believe in God, but I'm very interested in Her,' Clarke has famously said. Boardrooms in Europe and the United States are also the scene of a spiritual revival. Religion is said to be particularly strong among the new dotcom entrepreneurs, and California's Silicon Valley is host to a variety of alternative faiths.

Not that traditional church-going will fade. I have investigated for the *Guardian* newspaper an enormous and powerful evangelical Christian movement among middle-class students at the more 'Ivy league' British universities. It has routed student politics, which have virtually disappeared. An evening at a teetotal student party at the prestigious Durham University in 1999 suggested to me that a new and profoundly conservative and anti-intellectual elite may come to take an alarming slice of power in Britain and elsewhere this century.

It is quite disturbing to be informed by confident twenty-year-old science students that they regard evolution and the Big Bang as a fraud, or by a social work major that he has spent time on university helplines trying to 'cure' homosexuals. 'If God said the world was created in a week, then that's it for me,' said an English student. 'Is there just a chance that "a week" was a metaphor for people to understand the concept of billions of years?' I wondered. 'No,' she replied.

I touched in chapter four on a growing interest in medical research on the alleged benefits of prayer on sick patients, and I am sure this drug-free form of alternative therapy will be a further growth area this century. A poll of 1,004 Americans for *TIME* and CNN in 1996 found that 82 per cent believed in the healing power of prayer, and 64 per cent that doctors should pray with their patients. Whether prayer works for Divine reasons or as some kind of suggestive or self-hypnotic phenomenon will be debated, but the ascendancy of religion of various kinds, including meditation, will have a beneficial spin-off of making people feel physically healthier. As people talk about this, yet more will be attracted to spirituality. They will doubtless attribute their feeling of physical wellbeing to the rituals and forms of their chosen religious code rather than any possible placebo effect. This will be a cause of constant irritation and frustration to the majority of scientists.

It will be interesting, by comparison, to see how the scientific world reacts to any sign we get this century or this millennium that SETI – the search for extraterrestrial intelligence – has finally borne fruit. The public is certainly ready for ET. Whereas in 1600, a defrocked priest in Naples was burned at the stake for his belief that there might be other intelligences in other worlds, 54 per cent of Americans now believe there are aliens.

A minority of scientists remain unconvinced, believing that human existence is a one-off freak. Some, like Stephen Hawking, take a lateral view on the matter: 'There is a sick joke that the reason we have not been contacted by extraterrestrials is that when a civilization reaches our stage of development it becomes unstable and destroys itself,' he says. (He adds that he is nevertheless an optimist and thinks 'we have a good chance of avoiding both Armageddon and a new Dark Ages'.)

For the majority of scientists, however, the sheer size of the universe argues that not only do there have to be other intelligences, but that there really ought to be one somewhere where they speak English, too. Frank Drake, the radio astronomer who started the SETI movement in 1961, devised a mathematical formula suggesting that there are 10,000 technological civilizations in our galaxy alone. The late Carl Sagan, who was Professor of Astronomy and Space Sciences at Cornell University, and author of the novel *Contact*, believed there were a million intelligent civilizations in the Milky Way – which is, so far as we know, just one of 100 billion galaxies in the universe.

I would suspect, however, that even if there are so many intelligences out there, our chances of hearing from one of them is frustratingly low. Just one of the problems is timing; we would have to be exceedingly lucky to locate one where life is going on currently, and didn't happen millions of years ago, or be going to happen millions of years in the future. Additionally, the fact of being restricted to speed-of-light communications will, surely, reduce the chance of meaningful contact still further. The difficulties of being billions of years out of synch with ET will make the irritation of doing business with California from a Europe a mere eight hours ahead or Japan nine hours ahead seem a very minor one. Arthur C. Clarke, it has to be said, disagrees. 'I would bet [in the coming millennium] on the detection of alien life. It seems totally improbable that there isn't life beyond planet Earth, even if it only takes the form of microbes. Of course, the thought of intelligent life elsewhere is even more exciting – and more probable,' he told the *Irish Times* in an end-of-millennium interview.

What I would suggest is that news of extraterrestrial life, if it does come, could have a fascinating effect on humanity's internal disputes and rivalries, perhaps even causing us at last to act as one species. I had always thought this to be too naïve a thought to express, until I read this recent comment by Dame Margaret Anstee, a former under secretary of the United

Nations, and head of the UN peacekeeping mission to Angola in 1992–93. 'I am by nature optimistic, but I sometimes think the only solution will be an invasion from outer space,' Dame Margaret was saying. 'Then at last everyone would unite.'

What, then, of the prospects for electronic gadgetry, which, unlikely as it seems, really might hold centre stage in the future? Which are the technologies that will emulate TV and the cellular phone, and seize the public imagination to the point that they change society?

I strongly suspect that devices as intrusive as Orange's conceptual project, 'The Stud', with which I opened my preface, will largely be rejected by the public. Resentment at breaches of personal privacy is going to be a major theme of coming decades, and having an all-purpose, twenty-four-hour electronic assistant in one's ear will simply spook people out. I imagine an increasing number of people in the coming years will cut themselves off deliberately from the wired world. A suitably equipped spy can tell what you are typing on your personal computer by picking up radiation from the monitor unit from a block away; expect, therefore, a small boom in old-fashioned typewriters and fountain pens for ultra-personal communications.

As for slightly more modern technologies which will prove a hit, I would suggest a big 'no' to household robots – too clumsy when they don't work, way too creepy when they do – and only a qualified thumbs up to virtual reality. The received wisdom is that VR will become indistinguishable from RL (what VR enthusiasts call real life) when it comes to sex with virtual partners and exotic holidays without leaving home. But I wouldn't bet on it.

One technology that I think will have legs, however, is an extension of VR, known as tele-immersion. This is a proposed communications system designed to enable us to have the visual experience of being in the same room as someone who is actually in another city or country.

According to Jaron Lanier, the cult computer scientist, composer, visual artist, and author who invented the term *virtual reality*: 'It will just become part of life. It will be used by teenage girls to gossip, by business people to cut deals, by doctors to consult.' The heavily dreadlocked multi-millionaire Lanier, who grew up in a remote part of New Mexico in a geodesic-domed house with a concert pianist and a science writer for parents, is the lead scientist of the National Tele-immersion Initiative, a coalition of university departments studying applications for what is being called *Internet 2.*

The other key gizmo likely to be a twenty-first-century staple is the quantum computer, which will be hundreds of times faster than the speediest electrical machine. Current computers are restricted by the plodding speed at which electrons whiz around circuits. By making those circuits smaller, the process has been speeded up a little, but what is needed for the next century is a successor to electronics, which doesn't involve any connecting wires between components.

As in so many fields, the weirdness of quantum mechanics is expected to step in here, enabling electrons to jump from component to component instantaneously. Watch out for terms like 'quantum ratchet', the first and most basic of the new wave of quantum computer components which is already being developed. Palmtop computers faster than the largest Cray super-computer today, and with enough memory to *store* most of the current internet, might be expected to be in our pockets twenty or thirty years from now.

We will as a matter of course use these PalmPilot equivalents as an addition to our own brains. When you go to see your accountant, you will plug your mini computer into your brain and load a financial program. To play golf, you will select a spatial awareness option; for lunch with your in-laws, you will ask the computer to refresh your mind with details of all their fascinating, old-time reminiscences of Nirvana and the night somebody called Princess Diana died.

The great artificial intelligence question will, of course, become a core issue in life by then. A huge ethical argument will rage between those who insist that artificial intelligence is effectively human – and those who laugh at the notion, and at the dweebs' gullibility at being taken in by manipulative, but arguably still wholly mechanistic, robot minds. The media will greatly enjoy running stories about the absurdities of machine intelligence as they once took the opportunity to poke fun at racial minorities. Alleged instances of computers refusing to perform certain tasks because they are too boring will become a media staple.

As exotic as our early dabbling with artificial intelligence will be, our homes and offices will change disappointingly little in the foreseeable future. Attempts to update the office working environment to one of 'hotdesking', where workers grab whichever desk they like each morning, park their laptop and get to work, have been hated by workers, who unsurprisingly enjoy having a regular desk with pictures of their children on it. Large-scale remote working at home will also, I have a sneaking suspicion, be stillborn in large part. Many companies which have experimented with it have found workers hankering for social contact after a very short time.

Our homes, I expect, will also look remarkably similar in fifty or a hundred years' time to the way they do now. Perhaps because our houses represent our biggest financial and emotional investment, we are less willing to change them with the times than any other part of our lives. Even in a supposedly modernist world, we get unduly excited about 'period features' and pay above the odds for archaic house layouts, while most 'designs of the future' do not make it beyond architects' doodles. The only impetus for millions of people to graduate to reasonably modern furniture has been the Swedish store, IKEA, which has moved popular design on a stage or two by the simple device of making it cheap.

Amongst other likely future trends which have occurred to me during the writing of this book have been these:

- Every successive generation will continue to think of itself as uniquely special, and enjoy the drama of imagining it is present at an extraordinary turning point in history, either by being near the beginning of an era, or close to its apocalyptic end. It is a fundamental, but often overlooked, fact of life, however, that common things are common, and that the simplest explanation for anything is usually the truest. Conversely, it is unlikely that any given time or place is particularly 'special' – a theory which offends the ego-centric human spirit, and was unpopular when first aired by the astronomer Copernicus, who argued for the first time that the Earth is not at the centre of the Universe, but in a humdrum location somewhere to one side. The astrophysicist Professor Richard Gott of Princeton has spun this into an all-purpose theory that it is always more likely that any given time in human history is a boring, insignificant era than a particularly important one. I think on balance the twenty-first century *will* be a special time – but that won't stop people of the twenty-second century thinking their time is even more special.

- Ecological catastrophe will stubbornly refuse to materialize. If one single development guarantees this, it is that tourism officials in rain-sodden Manchester, England, held a conference in 2000 with an environmental group called Sustainability North-West, at which it was resolved to plant trees and build shelters in Manchester and Liverpool to protect people from the savage sunshine in the globally-warmed twenty-first-century future. However, even as the traditional rain and cold weather continue to keep such places green and wet, doomsayers will continue to maintain that the global warming theory was right all along. If the weather gets colder, they will blame global warming; if it stays the same, they will congratulate themselves on having staved off global warming.

- The population crisis will also never quite happen, although people will continue to argue that Malthus was right, even though Malthus was wrong.

- Some natural resources may run out, but ingenious ideas to replace them never will. If you look at a problem like home heating, humans have moved seamlessly from smoky, in-efficient open fires in the middle of their huts to the revolutionary idea of the chimney to Benjamin Franklin's idea of iron stoves with back boilers, to modern electric and gas heating, to solar heating, and so on. Each development works better and with greater energy efficiency than the last. To illustrate the infinite scope for innovative ideas, the economist Paul Romer, of Stanford University's Graduate Business School, has made the fascinating calculation that the number of distinct computer programs which could be written for the average gigabyte hard drive is one followed by 2.7 billion zeros. The number of seconds since the Big Bang, by contrast, is one followed by 17 zeros, the number of atoms in the universe, one plus 100 zeros.

- If a disaster such as a meteor strike occurs on populated land, it will be somewhere that only dirt poor people live. The same goes for some future mega air crash, just as terrible diseases always affect the poor in preference to the rich. If there is a God, it has always seemed to be exceptionally protective of innocent bystanders who happen to be middle class.

- The middle class will adopt an increasingly ascetic, gadget-free lifestyle – and come on rather superior about it. When they do buy gadgets, they will prize analog displays rather than digital, and upmarket manufacturers will pander to this trend. This is already happening. A couple of manufacturers have brought out VCRs with an old-fashioned clock in place of the standard digital display.

- 'Real', 'natural' everything will be valued more and more. In the eighteenth century, it was Rousseau denouncing the artificiality of civilization and yearning for the revival of a notion he had of 'the noble savage'. In the last century in Britain, it was real beer, cheese and organic vegetables. In coming years, it will be everything from travel to orgasms to shops – the latter as a revolt against e-commerce. Plentiful and perfect nano-engineered products will make hand-made things especially sought after. 'Conceivably the ultimate sign of prosperity in a nanotechnological future will be eating off dinner plates you made yourself, with lopsided handles and fingerprints visible in the glazed surfaces,' suggests Timothy Aeppel of the *Wall Street Journal*.

- There will be a growing need to teach people how to have face-to-face conversations. E-mail and voicemail have made it commonplace to have entirely asynchronous 'conversation', frequently across time zones. Not only is our e-mail persona often rather different from how we normally present ourselves, giving us the opportunity to create a new person, but we often imagine we have had a real conversation with someone, when we have not actually spoken to them for months or years. The grammar of normal conversation could become a forgotten art, precipitating the need for lessons in schools.

- A new premium will also be placed on the value of silence. The most advanced country in the world for mobile communications is Finland. Why this should be is mysterious. Some foreigners in Finland believe the Finns have always been desperate to be garrulous, but only if it doesn't have to be face-to-face. The archetypal modern Finnish scene today, as described by one foreigner, is of a gloomy man sitting alone with a beer in a gloomy bar – speaking gloomily into a phone. Finnish teenagers refer to mobile phones as *kanny*, which means 'extension of the hand'. The question arises, however, when looking at the many, diverse and constant ways Finns keep in touch with one another, is, will we eventually run out of things to say in our permanently logged-on society? With cellular-free zones already designated in Australian airports and on Virgin trains in the UK, I predict the development of silent, conversation-free zones in offices, restaurants and on public transport. Communication-free retreats will also become hugely popular.

- A similar premium will be placed on getting out of our homes and meeting people serendipitously, on the street, in the mall, the coffee shop and the supermarket. The transnational special interest groups so well serviced by the internet will begin to resemble gated communities as they take over increasing percentages of our life in real-seeming tele-immersion encounters. Getting out and bumping into real people at random will begin to seem very special.

- The letters 'www' will soon seem an amusingly antiquated throwback to the turn of the century, having long been replaced by something else, and will only be used by the elderly, along with the archaic 'dotcom'. However, the internet and its successors will continue for the next fifty or a hundred years to be thoroughly modern and take us by surprise with the breadth and imagination of the ways it can be used, from medical to educational to social. But I am convinced that consumer e-commerce is going to be something of a disappointment. It will settle down to become just one among many ways of buying, and never attract a real mass following. Companies selling to the public will only survive if they are big enough to absorb losses for years, and feel they need to keep up an internet presence as a flag-flying exercise. Business to business ('B2B') e-commerce will be much bigger and more important, but less high profile. In the meanwhile, the vast majority of the brash young, goatee-bearded hopefuls and sassy click-chicks that haunt the e-commerce quarters of Palo Alto, London's Brick Lane and the *Silicon Sentier* in Paris's old garment district, will become embittered middle-aged people, looking back on the insane optimism of the goldrush days of 1998 to 2005 with a mixture of nostalgia and loathing.

- Software glitches will continue for the foreseeable future to be an ever-growing problem. People will start to be killed by them, possibly in large numbers. One air disaster, the 1995 crash of an American Airlines Boeing 757 in Colombia, was blamed partly on software failure in a June 2000 court decision in Florida. We will get more frustrated by software when even programs devised by supposedly intelligent computers start to crash.

- The nuclear family living in its own 'little box' will continue to be the basis of all societies for hundreds, and probably thousands, of years. As people become wealthier, more content and live longer, however, their search for 'happiness' will, paradoxically, become more frantic. Divorce will become decreasingly stigmatized, and the majority of people will typically have several lengthy relationships and more than one family.

- Almost everybody will have several jobs at the same time. The newly graduated student working on a phone farm while doing a bit of bar work and a spot of web development is already the norm, and the multi-skill, multi-job culture will percolate through to include the majority of people of working age.

- Stephen Hawking says he is confident that by the end of the twenty-first century, we will have achieved the Holy Grail of science, an ultimate Theory of Everything, which would finally close the embarrassing gaps in mankind's understanding of the universe. I would suggest that as soon as we have a Theory of Everything, it will be discovered that there is yet another theory underlying it. Theoretical physicists will not go out of business this century for lack of anything left to discover.

- Lack of news in an increasingly bland and consensus-ruled world will lead to newspapers and magazines becoming more trivial by the year. People who remember real news will be amazed by the banality of the stories making front pages.

- Our view of the future will remain, for many decades yet, peculiarly stuck in a science fiction model dating from the mid twentieth-century. I have before me a cutting from the *Oxford Times* newspaper of 18 February 2000. Under the headline 'Hair we go into the twenty-first century', appears a picture of a female model with a severe expression very much like that of the 1939 *Vogue* male model on the cover of this book. The model is wearing a metal face guard which looks like a fencing accessory and is hooked over her ears. I would be prepared to guarantee that whatever avant garde fashions the current century and the next throw up, metal face guards hooked over people's ears will not be among them.

At the end of this journey, I think the most wonderful thing I have learned about the future is that there is so much of it, that there is always space for another original idea about it. Genetically 'improved' human beings living for ever, their minds permanently networked to provide a huge global consciousness . . . the ability to teleport instantaneously around the world . . . the hardware widely available to transform any base material containing carbon, hydrogen and oxygen into anything from hamburgers to doorknobs to replacement body parts.

All these have their proponents, as we have seen. Yet could the frame of reference that will lead humanity into the *fourth* millennium not be electronics, or nanotechnology or quantum mechanics, or genetic engineering, but something quite different that we have never heard of? Or could all these technologies have their role, but be overshadowed by something the majority of scientists currently regard as a non-starter or a joke?

Professor Brian Josephson of Cambridge University, joint winner of the 1973 Nobel Prize for Physics, agreed when we spoke that it is nearly impossible to visualize much of the future, but was prepared nonetheless to hazard a guess. Josephson, who is sufficiently eminent in physics to merit his own entry in some English dictionaries for 'The Josephson Effect' (it concerns the phenomena which occur when an electric current passes through a thin insulating layer between two superconducting substances), believes that within a few years, much of what we now call the paranormal will be routine science.

Was he thinking of any particular area of the paranormal – ESP, life after death, faith healing – as being ripe for de-paranormalization?

The formidably reticent professor hesitated as he considered the question. 'Well, all of them, really,' he finally replied. 'We'll have an understanding of the paranormal and science will have expanded. At a broad level, I would say we will understand the mind and its relationship to matter much better. We'll be able to do all sorts of things using the power of mind. And medicine might have been changed to use subtle energies. The French scientist Benveniste already has this futuristic vision that instead of a drug, in the future, you will download a waveform, and he's now being taken more seriously, improving his methodologies and apparently various people can now replicate some of his methods.

'As for something like communication with the dead, well I suppose so, but I think many people would find it's not very helpful to maintain connections and one should move on.'

Is Professor Josephson, then, optimistic or pessimistic about our prospects for, say, a thousand years hence? Another unsettling pause; Nobel prize-winners do not tend to rush to fill gaps in conversation for the sake of it.

'I think, umm, possibly optimistic, because many of our problems might have been dissolved by, well, some, hmm, higher force.'

An inspiring thought, perhaps, and certainly from an unexpected source. But in a way, if you want an intelligent but lateral slant on the future, I believe you might as well ask a humorist as a physicist.

Trust a comedy genius like Steve Martin, then, to have this inventive slant on futurology for a piece he wrote for *The New York Times* of 2 January 2000, entitled 'The Third Millennium: So Far So Good'.

Martin bemoaned the fact that by the end of this millennium, not only would he be dead, but all his stuff would be owned by somebody else. 'There will be people living in my house, wearing ridiculous hairdos, who will think of me and my age as hideously old-fashioned and moronically stupid, and who will look at our newspapers and see ads for clothes-storage shrink-wrap suction machines that will make them roar with laughter.

'On the other hand,' Martin concluded, 'it is comforting to note that these people will also be frighteningly stupid, sitting on their "sunflower" chairs, wearing their "wigwam" slippers and eating brain-enhancing toad power-pellets just as embarrassing as anything we ever sat on, wore or consumed. And perhaps you and I will be a few atoms in the raindrops that fall on them and ruin their day.'

But perhaps my favourite keynote speech for the future was made by the oddball architect Buckminster Fuller in the middle of the last century. Born in Massachusetts in 1895, 'Bucky' spent much of his life on a crusade to improve the human condition. As an optimist, the only way for him was forward, as it should be for all of us.

'Think about it,' he said. 'We are blessed with technology that would be indescribable to our forefathers. We have the where-withal, the know-it-all, to feed everybody, clothe everybody, give every human on Earth a chance. We now know what we could never have known before – that we have an option for all humanity to "make it" successfully on this planet and in this lifetime.'

NOTES

Preface: The Stud

p.2   The Stud project was described by Kenny Hirschhorn, group director of strategy, imagineering and futurology at Orange plc, in his 1999 paper, *Vision for a Wirefree Future Transforming the Way We Use the Phone.*

p.5   My 1997 interview with Professor John Hasted was for an earlier book, a biography of Uri Geller. Hasted is one of the scientists who support Geller's claims to be a paranormalist.

Chapter One: The Way We Weren't

p.17   The figures on international distribution of Starbucks outlets are from the company's website. They are obviously subject to change.

p.20   The provenance of the quotation from a 1950s home economics textbook is uncertain. It has been widely circulated for many years on the internet and is said to have been published in New Zealand. I also remember it circulating before the internet, which greatly increases my confidence in its authenticity.

p.21   Sextus Julius Frontinus, whose embarrassing view on inventions is widely quoted, was a first-century-AD aqueduct specialist, and author of the *De aquis urbis Romae* ('Concerning the Waters of the City of Rome'), who progressed to becoming governor of Britain. Charles H. Duell's identical thought more than 1,800 years later is also widely quoted and appears in *The Guinness Dictionary of Regrettable Quotations*. The original source is not specified, however. Neither is the name of the British Customs official who was so sceptical about the future of air transport.

p.22   Six years after they emerged in a *Nature* article, Professor Richard Gott's views were brilliantly explained in an interview by Timothy Ferris in *New Yorker* magazine, 12 July 1999.

p.27   The marvellous quotation from Joseph Glanvill was unearthed by my Latin master at school, P.J.C. Murray, in a delightful booklet he wrote, *The Belief in Progress* (see bibliography). Joseph Glanvill's book, in which his inspired passage of futurology appears, is *The Vanity of Dogmatizing*. It is in the Bodleian Library, Oxford.

p.30   Some of the information here on Mercier and the impact of the Montgolfier brothers is from I.F. Clarke's book, *The Pattern of Expectation, 1644–2001* (see bibliography).

p.34   Rachel Emma Silverman's article 'The Future is Now', which appeared in the *Wall Street Journal* on 1 January 2000, is the source of some of the early American futurological gems.

p.39   John Maynard Keynes's apparent prediction of e-commerce was unearthed by Paul Wallace in a splendidly sceptical piece on the subject in *New Statesman*, 21 February 2000.

p.43   The description of the 1939 New York World's Fair is from *Yesterday's Tomorrows*, by Corn and Horrigan (see bibliography). Some of my 1940s to 1960s American futurology material here and in chapter two came to light through a wonderful area of America On Line's own web pages by Eric Lefcowitz, called *Retro Future*.

Chapter Two: Is Futurology Bunk?

p.58   Stephen Hawking's point about time travel was made in an interview with Nigel Farndale of the *Sunday Telegraph*, 2 January 2000.

p.60   Ian Pearson's writings on the future can be read on his website, at *http://www.futures.org.uk/members/ipearson.htm*.

p.68   T.S. Ananthu's paper can be found at *http://www.mkgandhi-sarvodaya.org/techno.htm*.

Chapter Three: Global Warning

p.77   His Holiness was quoted in Matathia and Salzman's book, *Next* (see bibliography).

p.78 The Worldwatch Institute report on melting ice was itself reported in the UK by *The Week* magazine, 15 April 2000.

p.82 The US Geological Survey's view on the Yellowstone caldera appears at: *http://vulcan.wr.usgs.gov/Volcanoes/Yellowstone/OFR95–59/OFR95–59.html*.

p.85 The Anti Global Warming Petition Project can be found at *http://www.oism.org/pproject*.

p.87 A good article on acid lakes appears in *Dive*, the magazine of the British Sub Aqua Club, April 2000, by Frederick Ehrenstrom and Siski Green.

p.88 For more on Invasional Meltdown, see article in *Trends in Ecology and Evolution* (Canada) by Hugh McIsaac and Anthony Ricciardi, March 2000.

p.92 Many of the references in the section on the new Ice Age come from Adrian Berry's excellent if provocative book, *The Next 500 Years: Life in the Coming Millennium* (see bibliography).

p.93 A comprehensive article on anti-global warming academics appeared in *The New York Times*, by William K. Stevens, 29 February 2000.

p.94 A definitive article on the inaccuracy of most ecological doom-mongering, 'Plenty of Gloom', appeared in *The Economist*, 20 December 1997. See also an important article on the same theme by Ronald Bailey in *Reason*, *www.reason.com*, May 2000 issue.

p.95 The assertion that today's new cars put out only 5 per cent of the pollution of a 1970 model came from the influential energy expert Dan Yergin of the Boston and Washington DC consultancy Cambridge Energy Research Associates.

p.96 *TIME*'s article mentioning Kalundborg, by Ivan Amato, appeared on 8 November 1999. The chicken feathers recycling reference is from *Scientific American*, April 2000, by Diane Martindale.

p.97 Niles Eldredge's comments are contained in his article in *TIME*, 8 November 1999.

p.99 The work on anti-meteor precautions at Los Alamos and Lawrence Livermore national laboratories and elsewhere was reported by *TIME* on 10 April 2000.

p.106 For the German survey on differing international attitudes to global warming, see Bray and von Storch, *Climate Science: An Empirical Example of Postnormal Science*, Institute of Hydrophysics, GKSS Research Center, Geesthacht.

p.110 The figures on internet use in Ghana are from *The Times*, 18 April

2000, and on The Gambia, from *World Report*, published in the *Independent*, 8 April 2000.

Chapter Four: Don't Lose Your Head

p.111 Professor White's head transplant article was in *Scientific American*, September 1999.

p.112 The Cryonics Institute website is *http://www.cryonics.org/*.

p.116 News of Glasgow University's 'video pill' was reported by the *Daily Telegraph*, 24 February 2000, as was that of Professor Richard Vincent's news of 'smart' pacemakers.

p.117 Professor Richard Gregory's comment was in the *Guardian*, in an excellent piece, 'Born to be Wired', by Andy Beckett.

p.122 Professor W. French Anderson is quoted from his article in *Newsweek*, 27 December 1999.

p.126 Dr Ho set out the outline of her thoughts in more detail in the *Far Eastern Economic Review* (Hong Kong), on 27 January 2000.

p.127 Professor Richard Dawkins set out his reservations about the more extravagant genetic engineering claims in the London *Evening Standard*, 3 April 2000.

p.127 Professor Hawking's White House talk, *Science in the Next Millennium*, appears in full on the White House website, at: *http://www2.whitehouse.gov/Initiatives/Millennium/shawking.html*.

p.131 A selection of Dr Shmuel Halevi's articles in the *Journal of Chinese Medicine* appear at *http://www.acumedico.com/*.

Chapter Five: The Geek Shall Inherit the Earth

p.144 The *Towards a Science of Consciousness* convention was covered by Dan Falk of *United Press International*, who reported Christof von der Malsburg's and Stuart Hameroff's comments made there. Roger Penrose's view was reported by the *Observer* magazine on 2 January 2000.

p.145 There is more on Dr Glenn Doman's work, in *Hothouse People* by the author and Jane Walmsley (see bibliography).

p.145 A succinct and accessible account of directed neuroplasticity by Sharon Begley appeared in *Newsweek* on 1 January 2000.

p.146 For more on 'The Ten Per Cent Myth', see an article by that name on the The San Fernando Valley Folklore Society's Urban Legends Reference Pages, *www.snopes.com*. A paper by Dr Walsh, *Reverse Engineering the Human Brain,* was published in 2000 by the Royal Society, London.

p.147 Professor Damasio's article from which he is quoted, 'How the Brain Creates the Mind', was in *Scientific American,* December 1999. Professor Damasio is author of a fascinating book on consciousness, *The Feeling of What Happens* (see bibliography).

p.151 Risto Linturi's article appeared in the European edition of *Newsweek* on 27 December 1999. More of his ideas are on *www.linturi.fi.*

p.151 Professor Michio Kaku's comments are quoted from the Channel 4 TV series, *Predictions,* aired in the UK in late 1999. He is the author of *Visions* (see bibliography) and has the website *www.mkaku.org.*

p.152 The coming of The Grid, the successor to the internet, was reported by Clive Cookson, science editor of the *Financial Times,* 4 March 2000.

p.153 Hans Moravec's views are contained in his article 'Rise of the Robots', in *Scientific American,* December 1999. Professor Moravec's website is: *www.fre.ri.cmu.edu/users/hpm.* News of the Japanese robot Mei Mei was reported by the *Guardian* in its Online section, 25 May 2000.

p.156 The definitive article on Hans Peter Salzmann and the work being done with him at Tübingen University was 'Thinking Makes it So' by Ian Parker in the *Sunday Telegraph* magazine, 21 May 2000.

p.156 Professor Warwick was speaking at the British Association's *Festival of Science* in Sheffield in September 1999.

p.157 Ray Kurzweil's comments are in an article written by him for Amazon.com and appear among the reviews there for *The Age of Spiritual Machines.*

p.164 Sir Clive Sinclair's comment was in an article on David Potter, the founder of Psion, by Andy Beckett in the *Guardian* on 4 February 2000.

p.166 The holographic teacher experiment was reported by the *Express,* London, 15 February 2000.

Chapter Six: OHM Truths

p.170 Estimates of the likely level of e-commerce in the near future are from the author's own article, 'Dot Con?' in the *FT*'s 'How To Spend It', April 2000.

p.170 Ian Pearson's predictions of kitchen rage were reported by the *Independent*, 21 February 2000.

p.171 The internet microwave was mentioned in 'Can a Microwave Bank Cook the Books?' by Patrick Sherwen, the *Guardian*, 4 December 1999.

p.173 Howard Baetjer was quoted by Timothy Aeppel in the *Wall Street Journal*, 1 January 2000.

p.175 The anecdote about the DTI civil servant's trousers was told to the author by Professor David Gann of Sussex University, for a 1998 *FT* article on home automation.

Chapter Seven: Time Out

p.187 A good piece for further reading on this and a source for some of the material in this chapter is '2010, A Leisure Odyssey', by David Smith and John Harlow, *The Sunday Times*, 28 November 1999.

p.191 The 1982 Swedish study was quoted in Matathia and Salzman's book, *Next* (see bibliography).

p.193 For more on Missy the dog, see 'Make Me Another One', by James Langton, the *Sunday Telegraph* magazine, 9 April 2000 and news report by Guy Dennis in *The Sunday Times*, 7 May 2000.

p.195 'Watching the Cyber Athletes', by Steven Poole in the *Guardian*'s Online section, 25 May 2000, was a source for part of the section on computer games.

Chapter Eight: What's the Big Idea?

p.203 Mr Obuchi's obituaries were amusingly extracted in *The Week* magazine on 20 May 2000.

p.205 *Newsweek*'s December 1999 special edition, *Issues 2000,* is a good source for further reading on the future of politics.

p.207 Kenneth Baker was quoted in the London *Evening Standard* in 1988 in a report by Bruce Kemble.

p.209 Some of the examples of internet use by people in remote corners of the world are culled from 'A Wider Net', by Jennifer L. Schenker in *TIME*'s 11 October 1999 special, *The Communications Revolution*.

Chapter Nine: To Boldly Go

p.221 For more on the Moller Skycar, see 'The Sky's the Limit', by William Langley, the *Sunday Telegraph* magazine, 2 January 2000.

p.222 News of Visteon's car navigation system was reported by Ian Adcock in *The Times*, 9 October 1999.

p.222 Professor Newland was speaking to the Cambridge Alumni magazine, Autumn 1999 edition.

p.224 The autogiro taxi plans for Shanghai were reported by *T3* magazine (UK) in June 2000, in an excellent article by Paul McCauley on the future of transport, which was a source for parts of this chapter.

p.229 A fairly accessible piece on quantum teleportation by Anton Zeilinger, lately of the Institute for Experimental Physics at the University of Vienna, appeared as the cover story in *Scientific American*, April 2000.

p.230 Professor Gott's thoughts on time travel appeared in a *TIME* special on the future of space exploration, 10 April 2000.

p.234 Malcolm Walter is author of a 1999 book, *The Search for Life on Mars* (see bibliography).

Chapter Ten: Mystic Megabyte

p.243 An interesting piece by Andrew Moody on spirituality in the boardroom appeared in the *Mail on Sunday* financial section on 9 January 2000.

p.244 Hawking's view here is taken from his White House address (see p. 127 above).

p.245 Arthur C. Clarke's view on aliens was expressed to David Orr of the *Irish Times*, 3 January 2000.

p.245 Dame Margaret Anstee was writing in a special late 1999 millennium edition of the magazine for Cambridge University alumni.

p.250 Paul Romer's calculation appeared in *Worldlink*, January/February 1995, available at *http://www.stanford.edu//promer/wrld_lnk.htm.*

p.250 Philips is one among several manufacturers who have brought out retro VCRs with an analog clockface.

p.251 Some of the material on cell phones in Finland comes from an article by William J. Holstein in *U.S. News & World Report,* 13 December 1999.

p.253 For anybody who really wants to see it – and it is very funny – the hairstyles of the twenty-first-century photo were in the *Oxford Times* on 18 February 2000.

p.256 Buckminster Fuller's quote is taken from America On Line's *Retro Future* feature.

BIBLIOGRAPHY

Abrams, Malcolm and Harriet Bernstein, *Future Stuff: More than 250 Useful, Time-Saving, Delicious, Fun, Stimulating, and Energy-Saving Products that Will Be Available by the Year 2000*, Penguin, 1989.

Abrams, Malcolm and Harriet Bernstein, *More Future Stuff: Over 250 Inventions that Will Change Your Life by 2001*, Penguin, 1991.

Atkinson, Austen, *Impact Earth*, Virgin Publishing, 1999.

Azam, Dr Ikram, *Towards the Third Millennium: The Islamic Vision, World-View and Mind*, The Pakistan Futuristics Foundation and Institute, Islamabad, 1999.

Bagrit, Sir Leon, *The Age of Automation*, Weidenfeld & Nicolson, 1965.

Bell, Art and Whitley Strieber, *The Coming Global Superstorm*, Pocket Books, 1999.

Bernal, J.D., *The World, the Flesh and the Devil*, first pub. 1929, latest ed. Jonathan Cape, 1970, with author's new note.

Berry, Adrian, *The Next 500 Years: Life in the Coming Millennium*, Headline, 1995.

Berry, Adrian, *The Giant Leap: Mankind Heads for the Stars*, Headline, 1999.

Bramwell, Anna, *Ecology in the Twentieth Century: A History*, Yale University Press, 1989.

Casti, John L., *Searching for Certainty: What Scientists Can Know about the Future*, William Morrow, 1991.

Clark, John O.E., *Computers at Work*, Hamlyn, 1969.

Clarke, Arthur C., *July 20, 2019: A Day in the Life of the 21st Century*, Grafton Books, 1986.

Clarke, Arthur C., *2001: A Space Odyssey*, Orbit, 1990.

Clarke, I.F., *The Pattern of Expectation, 1644–2001*, Jonathan Cape, 1979.

Corn, Joseph J. and Brian Horrigan, *Yesterday's Tomorrows: Past Visions of the American Future*, Johns Hopkins University Press, 1996.

Damasio, Antonio, *The Feeling of What Happens*, Heinemann, 2000.

Evans, Christopher, *The Mighty Micro: The Impact of the Micro-Chip Revolution*, Victor Gollancz, 1979.

Fukuyama, Francis, *The End of History and the Last Man*, Penguin, 1993.

Gell-Mann, Murray, *The Quark and the Jaguar*, Abacus, 1995.

Gershenfeld, Neil, *When Things Start to Think*, Hodder & Stoughton, 1999.

Ed. Griffiths, Sian, *Predictions*, OUP, 1999.

Halperin, James L., *The Truth Machine: A Speculative Novel*, Simon & Schuster UK, 1997.

Halperin, James L., *The First Immortal: A Novel of the Future*, Ballantine, 1998.

Huxley, Aldous, *Brave New World*, Chatto & Windus Ltd, 1932.

Kaku, Michio, *Visions*, OUP, 1999.

Ed. Kelly, Catriona, *Utopias: Russian Modernist Texts 1905–1940*, Penguin, 1999.

King, Larry and Pat Piper, *Future Talk: Conversations about Tomorrow*, HarperCollins, 1998.

Kurzweil, Ray, *The Future Now: Predicting the Twenty-First Century*, Weidenfeld, 1998.

Kurzweil, Ray, *The Age of Spiritual Machines*, Orion Business, 1999.

Lacey, Robert and Danny Danziger, *The Year 1000: What Life Was Like at the Turn of the First Millennium*, Little, Brown, 1999.

Lee, Christopher, *This Sceptered Isle*, Penguin, 1998.

Lovelock, J.E., *Gaia: A New Look at Life on Earth*, OUP, 1979.

McGuire, Bill, *Apocalypse: A Natural History of Global Disasters*, Cassell, 1999.

Matathia, Ira and Marian Salzman, *Next: Trends for the Near Future*, The Overlook Press, 1999.

Miller, Geoffrey, *The Mating Mind*, Heinemann, 2000.

Murray, P.J.C., *The Belief in Progress*, Edward Arnold, 1971.

Orwell, George, *1984* (various editions).

Palfreman, Jon and Doron, Swade, *The Dream Machine: Exploring the Computer Age*, BBC Books, 1991.

Poole, Steven, *Trigger Happy: The Inner Life of Videogames*, Fourth Estate, 2000.

Sagan, Carl, *Contact: A Novel*, Century Hutchinson, 1986.

Schwartz, P., P. Leyden and J. Hyatt, *The Long Boom – A Vision for the Coming Age of Prosperity*, Nexus Special Interests, 1999.

Shlain, Leonard, *The Alphabet Versus the Goddess*, Penguin, 1999.

Toffler, Alvin, *Future Shock*, Pan, 1973.

Walmsley, Jane and Jonathan Margolis, *Hot House People: Can We Create Super Human Beings?*, Pan, 1987.

Walter, Malcolm, *The Search for Life on Mars*, Perseus Books, 1999.

Warwick, Kevin, *March of the Machines: Why the New Race of Robots Will Rule the World*, Century, 1997.

Wieners, Brad and David Pescovitz, *Reality Check*, Hardwired, 1996.

Wills, Christopher, *Children of Prometheus: The Accelerating Pace of Human Evolution*, Allen Lane, The Penguin Press, 1999.

FILMOGRAPHY

Metropolis, dir. Fritz Lang, 1926.
Frankenstein, dir. James Whale, 1931.
Fantastic Voyage, dir. Richard Fleischer, 1966.
2001, A Space Odyssey, dir. Stanley Kubrick, 1968.
Blade Runner, dir. Ridley Scott, 1982.
Cocoon, dir. Ron Howard, 1985.
Total Recall, dir. Paul Verhoeven, 1990.
Armageddon, dir. Tiber Takacs, 1997.

INDEX

A NOTE ON THE AUTHOR

Jonathan Margolis writes on new technology
and hi-tech consumer products for the *Evening
Standard*, the *Financial Times*, the *Daily Mail*,
GQ, *Elle*, and *Time* magazine. He has a regular
'new gadgets' slot on Sky TV News. He is the
author of a critical biography of Uri Geller
and (with Jane Walmsley) of *Hothouse People:
Can We Create Super Human Beings?*

A NOTE ON THE TYPE

The text of this book is set in Linotype Sabon,
named after the type founder, Jacques Sabon. It
was designed by Jan Tschichold and jointly
developed by Linotype, Monotype and Stempel,
in response to a need for a typeface to be
available in identical form for mechanical hot
metal composition and hand composition using
foundry type.

Tschichold based his design for Sabon roman on
a fount engraved by Garamond, and Sabon italic
on a fount by Granjon. It was first used in 1966
and has proved an enduring modern classic.